The Theory of Sheaves

CHICAGO LECTURES IN MATHEMATICS

The Theory of Sheaves

RICHARD G. SWAN

Chicago and London
THE UNIVERSITY OF CHICAGO PRESS

Library of Congress Catalog Card No. 64-24979
THE UNIVERSITY OF CHICAGO PRESS, CHICAGO 60637
The University of Chicago Press, Ltd., London W.C. 1

Published 1964. Second Impression 1968
Printed in United States of America

ACKNOWLEDGEMENT

These notes are based on a course of lectures given in Dr. I. M. James' seminar at Oxford in 1958. The original notes have been revised and rearranged by R. Brown and R. G. Swan; the latter has also added much material not in the lectures. We are grateful to the Mathematical Institute of the University of Oxford for permission to reproduce the notes here.

Much of the material was adapted from Seminaire Henri Cartan 1950-51 and A. Grothendieck - "Sur quelques points d'algebre homologique," Tohoku Math. Journal, August, 1957.

For help in editing the manuscript for publication, we should like to thank C. Procesi.

CONTENTS

INTRODUCTION

This introduction is intended primarily for those readers who have no previous knowledge of the theory of sheaves. Such readers probably have already asked themselves at least two questions:

(1) What are sheaves?

(2) What are they good for?

I will try to answer these questions here, beginning with question 2. The obvious answer to this question is that sheaves are very useful in proving theorems, for example:

<u>In topology</u>

1. The theorem that singular and Čech cohomology agree for paracompact HLC spaces.

2. The duality theorems of Poincaré, Alexander, and Lefschetz, and hence their consequences such as the Jordan-Brouwer theorem.

3. The existence of a spectral sequence associated with a map and consequences of this, such as the Vietoris mapping theorem and the spectral sequence of a fiber space.

<u>In differential and algebraic geometry</u>

4. The theorems of de Rham and Dolbeaut concerning cohomology groups defined by means of differential forms.

5. The duality theorem of Serre and its consequences (see (7)).

6. The Riemann-Roch theorem for varieties of all dimensions (see (6))

and many others.

Of course, many of these theorems can be proved without the use of sheaves. However, the proofs given by the theory of sheaves are generally much simpler and clearer than the classical proofs and hold under much more general conditions, especially the proofs of the Poincaré, Alexander, and Lefschetz duality theorems.

At first sight, the theorems listed above may not appear to have much in common. However, a closer examination of the theorems and their proofs reveals some basic similarities. We start by assuming (or proving) that a space has certain local properties. We then express these properties of the space in terms of the properties of certain sheaves over the space. Finally, we apply the theory of sheaves to deduce global properties of the space. Thus, as far as applications to topology are concerned, the importance of the theory of sheaves is simply that it gives relations (quite strong relations, in fact) between the local and global properties of a space. Before explaining the nature of these relations, it is necessary to return to question 1: "What are sheaves?"

A sheaf is essentially a system of local coefficients over a space X. In the usual cohomology theories, we consider a space X and a group G and define cohomology groups $H^n(X,G)$. In the theory of sheaves, we consider not a single group G, but a whole collection of groups G_x, one for each point $x \in X$. In the applications, of course, these groups will not be chosen arbitrarily but will be defined in some natural way in terms of properties of X. For convenience, we always regard the groups G_x as being disjoint. This can always be achieved by replacing each G_x by an isomorphic copy of G_x. The structure just defined is called a protosheaf[1] over X. The groups G_x are called the stalks of the protosheaf. We may picture a protosheaf as follows:

Let G be the union of all the G_x. Let $p: G \to X$ be defined by $p(g) = x$ if $g \in G_x$. Since the G_x are disjoint, this is well defined. Note that G is just a set with no topology so there is no question of continuity involved. This construction leads to the following picture,

---

[1]This term is due to Dr. I. M. James.

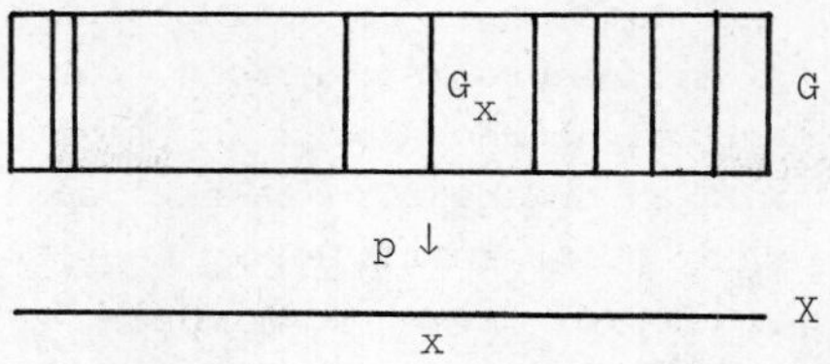

The vertical lines represent some of the stalks of G. The analogy with sheaves and stalks in the agricultural sense is quite apparent.

It is possible to express quite a bit of information about the local structure of X in terms of protosheaves. However, certain properties will be completely overlooked if we stick to using protosheaves with no additional structure. Namely, there is no relation between the different stalks of a protosheaf. This is best illustrated by means of examples.

The constant protosheaf

A very trivial example of a protosheaf is given by choosing some fixed group A and taking each G_x to be an isomorphic copy of A. In this protosheaf, any two stalks are canonically isomorphic. However, this is not apparent from the protosheaf structure itself. While it is obvious that two stalks are isomorphic, there will be many such isomorphisms in general and no reason to prefer one over another. The introduction of the canonical isomorphisms gives an additional element of structure.

The protosheaf of local homology groups

A more illuminating example is given as follows:

Let X be a topological n-manifold without boundary. Define a protosheaf G by $G_x = H_n(X, X - x)$ (regarding these groups as disjoint). By excision, $G_x \approx Z$ for all x. Therefore, this protosheaf is isomorphic to the constant protosheaf with stalk isomorphic to Z. However, we can again introduce an additional element of structure in G. Suppose we choose some point $x \in X$ and a generator $z \in H_n(X, X - x)$. Then z determines an orientation of X at x. Now, X may not be an orientable manifold, but in any case, every small enough neighborhood of x is orientable. Therefore, the choice of an orientation at x gives a unique orientation at every point sufficiently near to x. This then gives us a canonical isomorphism, $G_x \approx G_y$, for all y sufficiently close to x.

The local system formed by the cohomology of fibers

Let $f: E \to X$ be a fiber bundle. Let F_x be the fiber over x for each $x \in X$. Define a protosheaf G over X by $G_x = H^n(F_x)$ for some fixed n. As usual, we regard these groups as disjoint. Since $f: E \to X$ is a fiber bundle, each $x \in X$ has a neighborhood N_x such that $f^{-1}(N_x)$ is a product over N_x. This product structure then gives canonical isomorphisms $G_y \to G_x$ for all y near enough to x. (cf. Steenrod, Topology of Fibre Bundles, Part III, § 30.2).

In all the examples so far considered, the additional structure imposed on G reduces to this: G has a topology which makes it a bundle of coefficients over X in the sense of Steenrod, Fibre Bundles, Part III, § 30.1. Such a bundle of coefficients is a special case of a sheaf, called a locally constant (or locally simple, or locally trivial) sheaf. We now consider protosheaves with an additional element of structure which cannot be expressed in this way.

The sheaf of germs of holomorphic functions

Let X be an open set in the complex plane, or, more generally a complex manifold of any dimension. Let G_x be the set of all power series convergent in a neighborhood of x (the neighborhood depending on power series). As usual, regard the G_x as being disjoint. Then G is a protosheaf over X. Now, if $g \in G_x$, then g is a power series convergent in a neighborhood of x and therefore defines an analytic function in a neighborhood N_x of x. At each point y of this neighborhood N_x, we can expand this function in a power series, thus getting an element of G_y. Therefore, given an element $g \in G_x$, there are unique associated elements in all nearby stalks, that is, in all G_y such that $y \in N_x$.

Now, in this case, we cannot make G into a bundle of coefficients. The difficulty is that the neighborhood N_x depends on the element g. There is no single N_x which will work for all $g \in G_x$. However, we still have enough structure to define a topology on G. In a bundle of coefficients, an element $h \in G_y$ is close to an element $g \in G_x$ if and only if y is close to x and h is the element corresponding to y under the canonical isomorphism $G_y \approx G_x$.

In the sheaf of germs of holomorphic functions, there is no canonical isomorphism $G_y \approx G_x$, but there is still a relation which tells us when $h \in G_y$ is the element associated with $g \in G_x$. Therefore, we can still define a topology on G by defining a neighborhood of g to be the set of all elements associated with g and lying in stalks near g. In the particular sheaf under consideration here, this topology is simply that in which $g \in G_x$ and $h \in G_y$ are neighboring elements if and only if x and y are close and g and h represent the same analytic function. The neighborhoods just defined have an important property in common with the neighborhoods of elements in a bundle of coefficients. Namely, every small enough neighborhood of $g \in G_x$ meets every stalk G_y (with y near x) exactly once. Furthermore, if we have one such neighborhood U of g, every smaller neighborhood is obtained by taking only those elements of U which are in stalks G_y with y in some small neighborhood of x. These properties are best expressed as properties of the projection $p: G \to X$. They are equivalent to the statement that every element $g \in G_x$ has a neighborhood U which is mapped homeomorphically onto a neighborhood of x. Such a map p is called a local homeomorphism.

We have now found almost all the properties used to define a sheaf. The final one we need is the continuity of addition. This is a standard axiom used in defining almost all structures which involve both algebra and topology. Since addition is defined only in the stalks, the property must be stated as follows:

If $g,h \in G_x$, $g',h' \in G_y$, g is near g', and h is near h', then g + h is near g' + h'.

We can now define a sheaf of abelian groups. It consists of

(1) a protosheaf of abelian groups $p: G \to X$

and (2) a topology on G

such that

(a) p is a local homeomorphism

and (b) addition in G is continuous.

There is no reason to consider only sheaves of abelian groups. We may also consider sheaves of modules over a ring K. The definition is almost the same except that the stalks are assumed to be K-modules rather than abelian groups and multiplication by elements of K is assumed continuous. In other words, if $k \in K$, $g \in G_x$, $g' \in G_y$, and g is near g', then kg is near kg'.

Similarly, we may define sheaves of any sort of algebraic structures, e.g., rings, non-abelian groups, etc. We simply assume that all G_x have the given structure, that p is a local homeomorphism, and that all algebraic operations are continuous.

It is now possible to outline the method by which we obtain relations between the local and global properties of a space. We consider a sheaf F of chain or cochain complexes associated with X. We then take the group of sections $\Gamma(F)$ of F, that is, the group of continuous functions $s: X \to F$ such that ps = identity. This Γ (F) will be, essentially, one of the ordinary chain or cochain complexes of X. Its homology or cohomology H $(\Gamma(F))$ will give certain homology or cohomology groups of X. Now, instead of applying Γ and then H, we can take homology (or cohomology) immediately. This gives a sheaf H(F) of local homology (or cohomology) groups of X. The problem is to find relations between $H(F)$ and $H(\Gamma(F))$. This is a standard problem in homological algebra. It is solved by showing the existence of certain spectral sequences involving F, $H(F)$, and the derived functors of Γ. These derived functors of Γ play a central role in the theory of sheaves. If G is a sheaf over X, we define the ith cohomology groups $H^i(X,G)$ of X with coefficients in G to be the result of applying the ith right derived functor of Γ to G. It can be shown that, if X is paracompact and G is constant, these groups agree with the usual Cech cohomology groups of X.

There are two spectral sequences involving F. One relates $H(X,H(F))$ with a "hypercohomology group" of F. The other relates $H(H(\ ,F))$, and in particular $H(\Gamma(F))$, with this same hypercohomology group. In order to get useful relations, it is necessary to compute the hypercohomology group. This can be done by means of the second spectral sequence provided certain cohomology groups are trivial. The importance of this is such that a good deal of the general theory of sheaves will be concerned with finding conditions under which sheaves have trivial cohomology.

In conclusion, I will outline the main parts of the general theory of sheaves. We first define sheaves, maps, sections, and various functors which can be applied to sheaves and then give a general method for constructing sheaves. We also show that sheaves form an exact category in the sense of Buchsbaum (5). We then show that every sheaf is contained in an injective sheaf. As a consequence of this, we can apply the methods of homological algebra (injective resolution, derived functors, etc.) to the study of sheaves. As indicated above, we define the cohomology groups of X in terms of derived functors of Γ. It now becomes necessary to find general conditions under which these cohomology groups are trivial. This is the only part of the general theory which makes extensive use of geometrical arguments (covering theorems, paracompactness, etc.). Finally, we define the hypercohomology sequences and give applications to some of the theorems listed at the beginning of this introduction.

For other applications we refer the reader to the papers of S. S. Chern (1) and O. Zariski (2) and the bibliographies there.

I. ALGEBRAIC PRELIMINARIES

i) CATEGORIES AND FUNCTORS

A category $\mathcal{A}$ is given by the following:

1. A collection of objects A.
2. A set $M(A,B)$ for any two objects $A, B \in \mathcal{A}$. The elements $g \in M(A,B)$ will be called maps. We shall frequently write $g: A \to B$ for $g \in M(A,B)$.
3. A function $M(B,C) \times M(A,B) \to M(A,C)$ for each triple of objects $A,B,C \in \mathcal{A}$. The image of $\psi \times \varphi$ in $M(A,C)$ will be denoted by $\psi\varphi$, the composition of ψ and φ.

These terms must satisfy the following axioms:

I. If $\alpha: A \to B$, $\beta: B \to C$ and $\gamma: C \to D$, then $\gamma(\beta\alpha) = (\gamma\beta)\alpha$.

II. For each $A \in \mathcal{A}$, $\exists$ $i_A: A \to A$ such that if $\beta: B \to A$, $\gamma: A \to C$ and $i_A\beta = \beta$ then $\gamma\, i_A = \gamma$.

It is easy to verify that i_A is unique.

A covariant functor $T: \mathcal{A} \to \mathcal{B}$ is a set theoretic function and also a collection of functions $T: M(A,B) \to M(T(A),T(B))$ satisfying

(1) $T(\alpha\beta) = T(\alpha)T(\beta)$

and (2) $T(i_A) = i_T(A)$.

For two covariant functors $S,T: \mathcal{A} \to \mathcal{B}$ a natural transtormation $f: S \to T$ shall be a function f, defined on $\mathcal{A}$, such that $f(A) \in M(S(A),T(A))$ and such that the following diagram is commutative for any $A,B \in \mathcal{A}$ and map $g: A \to B$:

$$\begin{array}{ccc} S(A) & \xrightarrow{f(A)} & T(A) \\ {\scriptstyle S(g)}\downarrow & & \downarrow{\scriptstyle T(A)} \\ S(B) & \xrightarrow[f(B)]{} & T(B) \end{array}$$

For a contravariant functor, the only differences are that if $g: A \to B$, then $T(g) \in M(T(B),T(A))$ and $T(\alpha\beta) = T(\beta)T(\alpha)$; and a natural transformation $f: S \to T$ of two contravariant functors will have $f(A) \in M(S(A),T(A))$ and the following diagram commutative:

$$\begin{array}{ccc} S(A) & \xrightarrow{f(A)} & T(A) \\ {\scriptstyle S(g)}\uparrow & & \uparrow{\scriptstyle T(g)} \\ S(B) & \xrightarrow[f(B)]{} & T(B) \end{array}$$

A natural transformation is an equivalence if it has an inverse.

A category $\mathcal{A}$ shall be called a K-category where K is commutative ring with unit if it has a distinguished object 0, the zero object, and if it satisfies the following additional axioms:

III. $M(A,B)$ is a K-module (throughout these notes we shall assume that a K-module M is unitary: i.e., the unit of K acts as identity in M).

IV. the function $M(B,C) \times M(A,B) \to M(A,C)$ is a bilinear map of K-modules.

V. $M(0,0)$ is the zero module, also written 0.

In this case it is easy to verify that $M(0,A) = 0 = M(A,0)$ for any $A \in \mathcal{A}$.

If $\mathcal{A}$ and $\mathcal{B}$ are K-categories, we generally require that a functor $T: \mathcal{A} \to \mathcal{B}$ should give a K-homomorphism $T: M(A,B) \to M(T(A),T(B))$ (or $M(T(B),T(A))$ if T is contravariant). Such a functor will be called a K-functor or linear functor.

In a K-category $\mathcal{A}$ we could forget about the operations of K; or K may be Z, the ring of integers, and each module an abelian group. We shall then call $\mathcal{A}$ simply an additive category and the K-functors additive functors.

ii) UNIVERSAL MAPS

Let $\mathcal{A}$ and $\mathcal{B}$ be categories. Let us be given a certain class of maps from objects in $\mathcal{A}$ to objects in $\mathcal{B}$ called $\mathcal{A}\mathcal{B}$ maps, which satisfy

(1) if $f: A' \to A$ is in $\mathcal{A}$, $g: B \to B'$ is in $\mathcal{B}$, and $h: A \to B$ is an $\mathcal{A}\mathcal{B}$ map, then $hf: A' \to B$, $gh: A \to B'$ are defined and are $\mathcal{A}\mathcal{B}$ maps;

(2) if $f_1: A'' \to A'$, $f_2: A' \to A$ are in $\mathcal{A}$,
$g_1: B \to B'$, $g_2: B' \to B''$ are in $\mathcal{B}$, and
$h: A \to B$ is an $\mathcal{A}\mathcal{B}$ map,
then $h(f_2 f_1) = (hf_2)f_1$,
$(g_2 g_1)h = g_2(g_1 h)$, and
$g_1(hf_2) = (g_1 h)f_2$; and

(3) if f,g are identities, then $hf = h$, $gh = h$.

<u>Definition</u>: An $\mathcal{A}\mathcal{B}$ map $u: A \to B$ is <u>right-universal</u> if for any $f: A \to B'$ there exists a unique $h: B \to B'$ such that $hu = f$.

$$\begin{array}{ccc} & u & \\ A & \to & B \\ f & \searrow\swarrow & h \\ & B' & \end{array}$$

An $\mathcal{A}\mathcal{B}$ map $u: A \to B$ is left-universal if for any $f: A' \to B$, there exists a unique $h': A' \to A$ such that $uh' = f$.

$$\begin{array}{ccc} & u & \\ A & \to & B \\ h' \nwarrow & & \nearrow f \\ & A' & \end{array}$$

Suppose all objects of $\mathcal{A}$ have at least one associated right-universal map. Let T be a map from objects of $\mathcal{A}$ to objects of $\mathcal{B}$ constructed by choosing a universal map $u: A \to B$ for each $A \in \mathcal{A}$ and letting $T(A) = B$.

If $f: A' \to A$, we have the following diagram:

$$\begin{array}{ccc} A' & \xrightarrow{f} & A \\ u'\downarrow & & \downarrow u \\ T(A') & & T(A) \end{array}$$

u' is right-universal, so the map uf can be uniquely factored through T(A') to give a map $T(f):T(A') \to T(A)$. It is easily checked that, by the uniqueness property, T is a covariant functor.

Let the functor $S: \mathcal{A} \to \mathcal{B}$ be obtained by a similar choice of right-universal maps. Then the diagram

$$\begin{array}{ccc} & A & \\ {}^{u_1}\swarrow & & \searrow^{u_2} \\ S(A) & & T(A) \end{array}$$

can be completed with maps $\tau(A):S(A) \to T(A)$, $\mu(A):T(A) \to S(A)$ by the universal property of u_1, u_2 with $\tau u_1 = u_2$, $\mu u_2 = u_1$.

The diagram

$$\begin{array}{ccc} & A & \\ {}^{u_1}\swarrow & & \searrow^{u_1} \\ S(A) & \xrightarrow[\mu\tau]{} & S(A) \end{array}$$

is commutative since $\mu\tau u_1 = \mu u_2 = u_1$ and hence $\mu\tau$ = identity. Similarly, $\tau\mu$ = identity; i.e., S and T are naturally equivalent.

Suppose each object $B \in \mathcal{B}$ has an associated left-universal map $u:A \to B$ for some $A \in \mathcal{A}$. Choose a particular such A for each $B \in \mathcal{B}$ and define $T(B) = A$. Then the following diagram can, by the left-universal property, be uniquely completed and we define $T(f) = f':T(B') \to T(B)$. We get a covariant functor $T: \mathcal{B} \to \mathcal{A}$, and by an argument similar to the earlier one, we can show that any two functors derived in this way are naturally equivalent.

$$\begin{array}{ccc} T(B') & \xrightarrow{f'} & T(B) \\ u'\downarrow & & \downarrow u \\ B' & \xrightarrow[f]{} & B \end{array}$$

<u>Remark</u>: Suppose $\mathcal{A}$ and $\mathcal{B}$ are additive categories and that the given class of $\mathcal{A}\mathcal{B}$ maps satisfies the following additional hypotheses:

(4) if A is in $\mathcal{A}$, B in $\mathcal{B}$, the $\mathcal{A}\mathcal{B}$ maps $h: A \to B$ form an abelian group.

(5) if $f_1, f_2: A' \to A$, are in $\mathcal{A}$, $g_1, g_2: B \to B'$, are in $\mathcal{B}$, and $h: A \to B$ is an $\mathcal{A}\mathcal{B}$ map,

then $h(f_1 + f_2) = hf_1 + Hf_2$ and $(g_1 + g_2)h = g_1h + g_2h$.

In this case the functors earlier defined by right (or left) universal maps are easily seen to be additive functors because, with the above notations, if the diagrams

$$\begin{array}{ccc} A' & \xrightarrow{f_1} & A \\ u'\downarrow & & \downarrow u \\ T(A') & \xrightarrow{T(f_1)} & T(A) \end{array} \qquad \text{and} \qquad \begin{array}{ccc} A' & \xrightarrow{f_2} & A \\ u'\downarrow & & \downarrow u \\ T(A') & \xrightarrow{T(f_2)} & T(A) \end{array}$$

are commutative, then the following diagram

$$\begin{array}{ccc} A' & \xrightarrow{f_1 + f_2} & A \\ u'\downarrow & & \downarrow u \\ T(A') & \xrightarrow{T(f_1) + T(f_2)} & T(A) \end{array}$$

is commutative. Therefore, by the uniqueness property, $T(f_1 + f_2) = T(f_1) + T(f_2)$.

<u>Examples</u>:

1. <u>Direct sums</u>

Let $\mathcal{A}$ be the category of sets of groups $\{A_\alpha\}_{\alpha \in I}$ for some fixed indexing set I, in which the maps are sets of homorphisms $g_\alpha: A_\alpha \to A'_\alpha$. Let $\mathcal{B}$ be the category of groups. Let an $\mathcal{A}\mathcal{B}$ map be a set of maps $f_\alpha: A_\alpha \to B$, $B \in \mathcal{B}$.

Then the map $\{i_\alpha\} : \{A_\alpha\} \to \Sigma A_\alpha$, where the i_α are the injections into the direct sum, is a right-universal $\mathcal{A}\mathcal{B}$ map, the map h being defined by $h = \Sigma f_\alpha \circ p_\alpha$ where p_α is the projection $\Sigma A_\alpha \to A_\alpha$

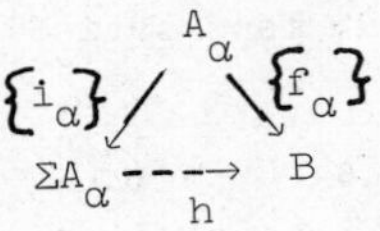

2. Direct products

Let the categories $\mathcal{A}$ and $\mathcal{B}$ be as in (1). A $\mathcal{B}\mathcal{A}$ map is now a set of homorphisms $g_\alpha: B \to A_\alpha$. The set of projections of the direct product $p_\alpha: \Pi A_\alpha \to A_\alpha$ is a left-universal $\mathcal{B}\mathcal{A}$ map. By the definition of the direct product, there is for any $x_\alpha \in A_\alpha$ a unique $x \in \Pi A_\alpha$ such that $p_\alpha x = x_\alpha$.

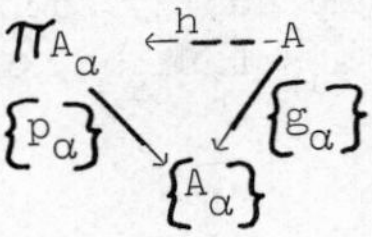

So the $\{g_\alpha\}$ can be "lifted" to a map $h: A \to \Pi A_\alpha$, which is easily seen to be a unique homomorphism.

3. Tensor products

Let $\mathcal{B}$ be the category of abelian groups and homomorphisms. Let $\mathcal{A}$ be the category whose objects are cartesian products A X B where A is a right Λ-module, B a left Λ-module, and whose maps are pairs (φ, ψ) of Λ-homomorphisms where $\varphi: A \to A'$, $\psi: B \to B'$. An $\mathcal{A}\mathcal{B}$ map shall be a bilinear map $f: A \times B \to C$; i.e., $f(a_1+a_2,b) = f(a_1,b) + f(a_2,b)$, $f(a,b_1+b_2) = f(a,b_1) + f(a,b_2)$, and $f(a\lambda,b) = f(a,\lambda b)$.

Theorem 1: Every object in $\mathcal{A}$ has an associated right-universal map.

I do not give the proof. The corresponding group we write $A \otimes_\Lambda B$, the tensor product of A and B; it can be taken to be the factor group P/R where P is the free abelian group generated by pairs (a,b) $a \in A$, $B \in B$ and R is generaged to relations

$(a,b) + (a,b') - (a,b+b')$, where $a,a' \in A$;

$(a,b) + (a',b) - (a+a',b)$, where $b,b' \in B$; and

$(a\tau,b) - (a,\tau b)$, where $\tau \in \Lambda$.

The universal map $u: A \times B \to A \otimes_{\Lambda} B$ takes (a,b) into the coset of (a,b) mod. R.

4. The group $\mathrm{Hom}_{\Lambda}(A,B)$

Let $\mathcal{A}$ be the category of abelian groups; let $\mathcal{M}$ be the category of left Λ-modules and Λ-homomorphisms. Let $\mathcal{M}^*$ be the dual category of $\mathcal{M}$; i.e., $\mathcal{M}^*$ has the same objects as $\mathcal{M}$ and its maps are the maps of $\mathcal{M}$, only written in the opposite direction. The product category $\mathcal{B} = \mathcal{M}^* \times \mathcal{M}$ has objects pairs of objects from $\mathcal{M}^*$ and $\mathcal{M}$ maps pairs $(\varphi,\psi):(A,B) \to (A',B')$, where $\varphi: A \to A'$, $\psi: B \to B'$ are maps of $\mathcal{M}^*$ and $\mathcal{M}$, respectively. φ is in fact a Λ-homomorphism $A' \to A$. By an $\mathcal{A}\mathcal{B}$ map $C \to (A,B)$ we mean a bilinear map $C \times A \to B$.

Theorem 2: Each object of $\mathcal{B}$ has an associated left-universal map.

The corresponding group we write as $\mathrm{Hom}_{\Lambda}(A,B)$. As a representative we can take the group, say $\chi(A,B)$, of Λ-homomorphisms $A \to B$, with the map $u: \chi(A,B) \times A \to B$ defined by $u(g,a) = g(a)$.

5. Direct limits

Definition: An ordered set is a double $(I,<)$ where $<$ is a transitive and reflexive relation on the set I. A map of ordered sets $f:(I,<) \to (J,<)$, often written $f: I \to J$, is a function $f: I \to J$ such that $\alpha < \beta \to f(\alpha) < f(\beta)$.

Any subset $J \subset I$ has a natural ordering induced by that of I, making $J \to I$, by injection, a map of ordered sets.

The product $(I,<) \times (J,<)$, written $I \times J$, is $(I \times J, <<)$ where $(\alpha,\beta) << (\alpha',\beta')$ if and only if $\alpha < \alpha'$ and $\beta < \beta'$.

An ordered set $(I,<)$ is directed if for any $\alpha,\beta \in I$, $\exists \gamma \in I$ such that $\gamma < \alpha$ and $\gamma < \beta$. As an example of a directed set, we may take the neighborhoods of a point x of a topological space, ordered by inclusion.

Definition: A direct system of K-modules is a directed set I; for each $\alpha \in I$, a K-module M_α; and for each two elements $\alpha, \beta \in I$ with $\alpha < \beta$ a K-homomorphism $\varphi_\alpha^\beta : M_\beta \to M_\alpha$ satisfying

φ_α^α = identity, $\varphi_\alpha^\beta \varphi_\beta^\gamma = \varphi_\alpha^\gamma$ if $\alpha < \beta < \gamma$.

A map of direct systems $(M_\alpha, I) \to (N_\beta, J)$ is

(i) a map $\sigma: I \to J$ of order sets and

(ii) for each $\alpha \in I$, a K-homomorphism $f_\alpha: M_\alpha \to N_{\sigma(\alpha)}$

so that for each two elements $\alpha, \beta \in I$ with $\alpha < \beta$, the following diagram commutes.

$$\begin{array}{ccc} M_\beta & \xrightarrow{f_\beta} & N_{\sigma(\beta)} \\ \varphi_\alpha^\beta \downarrow & & \downarrow \varphi_{\sigma(\alpha)}^{\sigma(\beta)} \\ M_\alpha & \xrightarrow{f_\alpha} & M_{\sigma(\alpha)} \end{array}$$

Note that if $J \subset I$, $M'_\alpha \subset M_\alpha$ for each $\alpha \in I$, the natural inclusion $(M'_\alpha, J) \subset (M_\alpha, I)$ is a map of direct systems, if and only if when $\beta < \alpha$, $\varphi'^\beta_\alpha : M'_\beta \to M'_\alpha$ is the restriction to M'_β of $\varphi_\alpha^\beta : M_\beta \to M_\alpha$.

Let $\mathcal{A}$ be the category of direct systems of K-modules and $\mathcal{B}$ the category of K-modules. By an $\mathcal{A}\mathcal{B}$ map $(M_\alpha, I) \to M$ we mean for each $\alpha \in I$, a K-homomorphism $f_\alpha: M_\alpha \to M$ such that if $\beta < \alpha$, $f_\alpha = f_\beta \varphi_\beta^\alpha$

$$\begin{array}{ccc} M_\alpha & \xrightarrow{\varphi_\beta^\alpha} & M_\beta \\ & f_\alpha \searrow \quad \swarrow f_\beta & \\ & M & \end{array}$$

<u>Theorem 3</u>: Each object in $\mathcal{A}$ has an associated right-universal map.

The associated functor is known as the direct limit of the direct system, and is written $\lim\limits_{\to I} M_\alpha$.

<u>Proof</u>: Without loss of generality we may suppose the M_α disjoint. Suppose $x_\alpha \in M_\alpha$, $y_\beta \in M_\beta$. We define an equivalence relation ~ in $\bigcup M_\alpha$ by $x_\alpha \sim y_\beta \leftrightarrow \exists \gamma \in I$ such that $\gamma < \alpha$, β and $\varphi_\gamma^\alpha(x_\alpha) = \varphi_\gamma^\beta(y_\beta)$. Let M = set of equivalence classes. Let $(x_\alpha), (y_\beta) \in M$. Then $\exists\, \gamma < \alpha, \beta$. Let $x_\gamma = \varphi_\gamma^\alpha x_\alpha$, $y_\gamma = \varphi_\gamma^\beta y_\beta$. Then $x_\gamma, y_\gamma \in M_\gamma$ and $(x_\gamma) = (x_\alpha)$, $(y_\gamma) = (y_\beta)$.

Define $\tau (x_\alpha) + \mu (y_\beta) = (\tau x_\gamma + \mu y_\gamma)$, $\tau, \mu \in K$. In this way, we give M the structure of a K-module. Let $u_\alpha : M_\alpha \to M$ be the natural map. Then $u = \{u_\alpha\} : (M_\alpha, I) \to M$ is an $\mathcal{AB}$ map. I assert

(1) u is epi. For if $x \in M$, $\exists x_\alpha$ such that $u_\alpha(x_\alpha) = x$ and

(2) u is "as mono as possible;" i.e., if $u_\alpha(x_\alpha) = 0$, then $\exists \beta$, $\beta < \alpha$, such that $\varphi^\alpha_\beta(x_\alpha) = 0$.

Suppose now we have an $\mathcal{AB}$ map $(M_\alpha, I) \xrightarrow{f} N$. Then by (2) ker. $u \subset$ ker. f, and so, since u is epi., we can uniquely factor f through u. We could define and prove existence in almost exactly the same way of direct systems of sets, i.e., assuming no algebraic structures in M_α.

<u>Definition</u>: Let $(I, <)$ be a directed set. $J \subset I$ is <u>cofinal</u> if for any $a \in I$, $\exists \beta \in J$ such that $\beta < \alpha$.

<u>Theorem 4</u>: Let (M_α, I) be a direct system, J cofinal in I. Then the inclusion map $(M_\alpha, J) \to (M_\alpha, I)$ induces an isomorphism $M_J = \varinjlim_J M_\alpha \approx \varinjlim_I M_\alpha = M_I$.

<u>Proof</u>: Let $i : M_J \to M_I$ be the induced map. Let $(x_\alpha) \in M_I$. Then $\exists \beta \in J$ with $\beta < \alpha$; let $x_\beta = \varphi^\alpha_\beta x_\alpha$, so that $(x_\alpha) = (x_\beta)$. Define $j : M_I \to M_J$ by $j(x_\alpha) = (x_\beta) \in M_J$. Then j is well defined, ji = identity, and ij = identity.

<u>Definition</u>: A sequence $\to (M'_\alpha, I) \to (M_\alpha, I) \to M''_\alpha, I) \to$ is <u>exact</u> at (M_α, I) if the two maps $I \to I$ shown are the identity and $M'_\alpha \to M_\alpha \to M''_\alpha$ is exact at M_α for all α.

<u>Theorem 5</u>: $\varinjlim$ preserves exactness.

Let $(M'_\alpha, I) \xrightarrow{i} (M_\alpha, I) \xrightarrow{j} (M''_\alpha, I)$ be exact. The theorem asserts that $M' \xrightarrow{\bar{i}} M \xrightarrow{\bar{j}} M''$ induced by $\varinjlim_I$ is exact.

Let $(x_\alpha) \in$ ker. j . Then $\exists x_\beta$ (x_α) such that $jx_\beta = 0$. Therefore $\exists y_\beta \in M'_\beta$ such that $iy_\beta = x_\beta$. Then $\bar{i}\ (y_\beta) = (x_\beta) = (x_\alpha)$.

Let $(x_\alpha) \in$ Im. $\bar{i}$. Then $\exists x_\beta \in (x_\alpha)$ such that $x_\beta = iy_\beta$. Therefore $\bar{j}\ (x_\alpha) = \bar{j}\ (x_\beta) = (jiy_\beta) = 0$.

<u>Theorem 6</u>: $\varinjlim$ commutes with $\otimes$; i.e., let $u : (M_\alpha, I) \to M$, $v : (N_\alpha, J) \to N$ be the universal maps.

$(M_\alpha \otimes N_\alpha, I \times J)$ is also a direct system. The theorem asserts that $\varinjlim_{I \times J} M_\alpha \otimes N_\beta \approx M \otimes N$ and that the corresponding maps $M_\alpha \otimes N_\beta \to M \otimes N$ are $u_\alpha \otimes v_\beta$.

To prove this, consider the maps $(x_\alpha) \otimes (y_\beta) \underset{j}{\overset{i}{\leftrightarrows}} (x_\alpha \otimes y_\beta)$, show that they are well defined, and prove ij = identity, ji = identity.

For later use, we mention the following: let M be a K-module and I the class of finite subsets of M, ordered by inclusion. For $\alpha \in I$, M_α shall be the submodule of M generated by elements of α. The inclusions $i_\alpha : M_\alpha \to M$ give a natural isomorphism $\varinjlim M_\alpha \approx M$.

Finally, to end this chapter, let us state a few theorems about $\otimes$.

<u>Theorem 7</u>: $\otimes$ is right exact; i.e., if $A' \xrightarrow{i} A \xrightarrow{j} A'' \to 0$ is exact, then so, for any B, is

$$A' \otimes B \xrightarrow{i \otimes 1} A \otimes B \xrightarrow{j \otimes 1} A'' \otimes B \to 0.$$

Let $a'' \otimes b \in A'' \otimes B$. Since j is onto, we have $a'' = ja$. Therefore, $j \otimes 1(a \otimes b) = a'' \otimes b$, and $j \otimes 1$ is onto.

Now $(j \otimes 1)(i \otimes 1) = (ji \otimes 1) = 0$, so we have a unique homomorphism u:coker. $(i \otimes 1) \to A'' \otimes B$. We prove this is an isomorphism. Let a bilinear map $g: A'' \times B \to$ coker. $(i \otimes 1)$ be defined by $g(a'', b)$ is the image of $a \otimes b$ in coker. $(i \otimes 1)$, where $ja = a''$; this is independent of the choice of a. Then $\exists$ a unique homomorphism $v: A'' \otimes B \to$ coker. $(i \otimes 1)$, such that $v(a'' \otimes b) = g(a'', b)$. Obviously vu and uv are identities.

We say that the exact sequence $0 \to A' \xrightarrow{i} A \xrightarrow{j} A'' \to 0$ <u>splits</u> if $\exists\, k: A'' \to A$ such that jk = identity. In this case, Im. k is a direct summand of A, the other summand being Im. i. Hence there is a map $h: A \to A'$ which is 0 on Im. k and with hi=identity. Finally, kj + ih = identity, and the sequence

$$0 \to A' \underset{h}{\overset{i}{\leftrightarrows}} A \underset{k}{\overset{j}{\leftrightarrows}} A'' \to 0$$

is exact both ways and splits.

Corollary: If the exact sequence $0 \to A' \to A \to A'' \to 0$ splits, then $0 \to A' \otimes B \to A \otimes B \to A'' \otimes B \to 0$ is exact and splits.

Suppose now K is a principal ideal domain. This implies that any finitely generated K-module is the direct sum of a finite number of K-modules with one generator. (cf. Bourbaki, Algebra.)

Definition: A K-module M is torsion-free if for $\lambda \in K$, $x \in M$, $\lambda x = 0 \to \lambda = 0$ or $x = 0$.

Obviously a submodule of a torsion-free module is torsion-free. Also if K is a principal ideal domain (as we assume from now on), then any finitely generated torsion-free K-module is free, i.e., is isomorphic to the direct sum of copies of K.

Theorem 8: If $0 \to A' \to A \to A'' \to 0$ is exact and B is torsion-free, then $0 \to A' \otimes B \to A \otimes B \to A'' \otimes B \to 0$ is exact.

B is the direct limit of finitely generated submodules B_α; each $B_\alpha \approx \sum_i^h K$. Therefore $M \otimes_K B_\alpha \approx \sum_i^h M$; i.e., the sequence $0 \to A' \otimes B_\alpha \to A \otimes B_\alpha \to A'' \otimes B_\alpha \to 0$ is the direct sum of copies of $0 \to A' \to A \to A'' \to 0$ and so is exact. Since $\otimes$ commutes with direct limits, the theorem is proved.

Theorem 9: If $0 \to A' \xrightarrow{i} A \xrightarrow{j} A'' \to 0$ is exact, A" torsion free and B arbitrary, then $0 \to A' \otimes B \to A \otimes B \to A'' \otimes B \to 0$ is exact.

Again write $A'' = \lim A''_\alpha$ where A''_α is finitely generated. Then $0 \to A' \xrightarrow{i} j^{-1}(A''_\alpha) \to A''_\alpha \to 0$ is exact. A''_α is free, so the sequence splits. Hence its exactness is preserved under tensor products. Proceeding to direct limits, we obtain the theorem.

Corollary: If $\to A_n \xrightarrow{i_n} A_{n-1} \xrightarrow{i_{n-1}} A_{n-2} \to \ldots$ is exact and A_i torsion-free, all i, then

$$\to A_n \otimes B \xrightarrow{i_n \otimes 1} A_{n-1} \otimes B \xrightarrow{i_{n-1} \otimes 1} A_{n-2} \otimes B \to \ldots$$

is exact for arbitrary B.

II. SHEAVES

§1

A sheaf is a triple (S,X,p) where S, the "sheaf", and X, the "base space", are topological spaces, and p, the "projection", is a continuous, onto map $S \to X$. (We shall often say: e.g., "let S be a sheaf" instead of the full notation "let (S,X,p) be a sheaf"). This shall have the additional structure:

i) TOPOLOGICAL

p is a local homeomorphism;

i.e., any $x \in S$ has a neighborhood U such that $p|U$ maps U homeomorphically onto a neighborhood of p(x).

This implies p is an open map.

For the rest of the structure, we need some definitions.

<u>Definition</u>: The <u>stalk</u> over $x \in X$ is the discrete set $S_x = p^{-1}(x)$.

<u>Definition</u>: Given two sheaves (S,X,p), (T,X,p^1), let S + T $= \{(x,y) \in S \times T;\ p(x) = p^1(y)\}$.

Define p": $S + T \to X$ by $p''(x,y) = p(x) = p^1(y)$.

Then $(S + T)_x = S_x \times T_x$.

ii) ALGEGRAIC

For the algebraic structure we suppose given once and for all a commutative ring K with unit. Then:

a. Assume each stalk S_x is given the structure of a K-module.

b. Each $\lambda \in K$ gives a map $\overline{\lambda}: S \to S$ by the obvious multiplication $\lambda : S_x \to S_x$. The map $S_x \times S_x \to S_x$ given by $(s_1, s_2) \to s_1 + s_2$.

gives a map $S + S \to S$.

We require both of these maps to be continuous.

<u>Definition</u>: A triple (S,X,p) in which S is a set, p is a

function $S \xrightarrow{\text{onto}} X$, and each $S_x = p^{-1}(x)$ has the structure of a K-module is called a <u>protosheaf</u>. Given any sheaf S we can form a protosheaf, written $\overline{S}$, by forgetting about the topology on S.

Much of the discussion to follow, e.g., difinition of maps of sheaves, restrictions, extensions, tensor products, etc. applies in an even simpler manner to protosheaves. All we do is omit all considerations of topology.

Note that (S + T,X,p") is a sheaf. $(\overline{S + T}, X, p")$ is a protosheaf. $\lambda : S + T \to S + T$ defined by $\lambda(a,b) = (\lambda a, \lambda b)$ is continuous and so is the composition

$$(S+T) + (S+T) \to (S+S) + (T+T) \to S + T$$

which is the addition in S + T. p" is obviously a local homeomorphism.

<u>Examples</u>:

i. <u>Constant sheaf</u>

Let M be a K-module. Give M the discrete topology, let S = X x M and let $p:S \to X$ be the projection $p(x,m) = x$.

The stalk over $y \in X$ is simply y x M.

This sheaf is usually denoted by M.

The constant sheaf of the zero module is the "zero sheaf", written O.

ii. <u>Bundle of coefficients</u>

<u>Definition</u>: A bundle of coefficients is a bundle of groups where the fibre is an abelian group and the group of the bundle is totally disconnected. This is a sheaf of Z-modules.

As an example let $X = S^1$, $G = Z_3$ and let the sheaf F be the union of a circle A and the double covering of S by a circle B. Each element of A shall be the zero in its stalk.

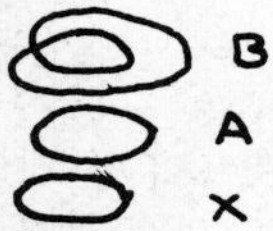

<u>Definition</u>: A map $\varphi:(F,X,p) \to (G,X,p')$ of sheaves (usually written $\varphi:F \to G$) is a continuous function $\varphi:F \to G$ such that (i) $p = p' \circ \varphi$ and (ii) $\varphi_x:F_x \to G_x$ is a K-homomorphism,

each $x \in X$ where $\varphi_x \overset{\text{dfn.}}{=} \varphi \mid F_x$.

$$\begin{array}{ccc} F & \rightarrow & G \\ & {}_{p}\searrow \quad \swarrow_{p'} & \\ & X & \end{array}$$

N.B. Any sheaf map $\varphi:F \rightarrow G$ is an open map of the topological spaces F and G.

Definition: A map $\varphi:F \rightarrow G$ is an isomorphism (written $\varphi:F \approx G$) if it is a homeomorphism of the topological spaces, F and G (Note: If φ_x is 1-1 and onto, it is certainly an isomorphism).

Definition: A sheaf F is trivial if it is isomorphic to the constant sheaf.

If $\varphi:F \rightarrow G$, $\psi:L \rightarrow M$ are sheaf maps, we have a sheaf map $\varphi \times \psi:F + L \rightarrow G + M$. We can define addition between maps $\varphi,\psi:F \rightarrow G$; $\varphi + \psi:F \rightarrow G$ is given by $(\varphi + \psi)_x = \varphi_x + \psi_x$.

For any $k \in K$, we can also define $k\varphi:F \rightarrow G$ by $(k\varphi)_x = k\varphi_x$. The set of sheaf maps $F \rightarrow G$ is a K-module written Hom (F,G). Hom (,) is easily seen to be a left-exact functor from sheaves to modules. So we can define the K-category $\mathcal{F}$ of sheaves and maps of sheaves. We shall be interested in constructing a homology theory for this category $\mathcal{F}$. The construction of this homology theory will follow the well-worn lines of homological algebra. Such a homology could be described axiomatically, and its uniqueness proved, on any exact category; i.e., an additive category in which, roughly speaking, we have (i) kernels, images, cokernels, and coimages of maps and (ii) direct sums of two objects.[1] Cartan's original proof of the uniqueness of cohomology theory of sheaves used essentially an additive category satisfying (i). Our proof (due to Grothendieck) will be somewhat simpler by using direct sums; though stated for sheaves, it can be generalized immediately to any exact category. The construction (also due to Grothendieck) will be both more general and simpler than Cartan's, mainly by using the tools of homological

[1] For more details see (5).

algebra as developed by Cartan and Eilenberg.

<u>Definition</u>: Let (G,X,p), (F,X,p') be sheaves. If $G \subset F$ as a set and this inclusion map is a sheaf map, we say (G,X,p) is a subsheaf of (F,X,p') (often written simply $G \subset F$).

<u>Proposition 1</u>: Let G be a subset of F where (F,X,p) is a sheaf. Then there is at most one sheaf structure on G making G a subsheaf of F. Such a structure exists if (1) $pG = X$; (2) G is open in F; and (3) G_x is a submodule of F_x, all $x \in X$.

We leave the proof to the reader.

<u>Example</u>: (iv) The sheaf F_U, where U is an open set in X and (F,X,p) is a sheaf, is defined by $(F_U)_x = \begin{cases} F_x & \text{if } x \in U \\ 0 & \text{if } x \notin U. \end{cases}$

<u>Quotient sheaf</u>: Let F' be a subsheaf of F. Form a protosheaf $\overline{F}/\overline{F'}$ with projection p" by letting $(\overline{F}/\overline{F'})_x = F_x/F'_x$. Define a map $\varphi : F \to \overline{F}/\overline{F'}$ by letting $\varphi \mid F_x$ be the natural map $F_x \to F_x/F'_x$; let F/F' be the protosheaf $\overline{F}/\overline{F'}$, together with the identification topology given by φ. The diagram

$$\begin{array}{ccc} F & \xrightarrow{\varphi} & F/F' \\ & {\scriptstyle p}\searrow \quad \swarrow{\scriptstyle p''} & \\ & X & \end{array}$$

is commutative.

<u>Proposition 2</u>: $(F/F', X, p'')$ is a sheaf; and φ is a sheaf map.

<u>Proof</u>: $y \in F/F'$, $\varphi(z) = y$, and $p(z) = x$. Let U be a neiborhood of z which is mapped homeomorphically by p onto a neighborhood V of x. Since φ is an open map, $\varphi(U)$ is a neighborhood of y. But $\varphi \mid U$ is 1-1, for no two elements of U belong to the same stalk, and so $\varphi \mid U$ is a homeomorphism. Therefore p" maps $\varphi(U)$ homeomorphically onto V, a neighborhood of x.

Finally we prove that addition and multiplication by $\lambda \in K$ in F/F' are continuous. To do this we need only consider the following commutative diagrams in which φ and $\varphi \times \varphi$ are sheaf epimorphisms, and so identification maps. (For any spaces Y,Z a map $Y \to Z$ which is a continuous, open onto map is an identification map.)

$$F/_{F'} + F/_{F'} \to F/_{F'} \qquad F/_{F'} \xrightarrow{\lambda} F/_{F'}$$
$$\varphi+\varphi \uparrow \qquad \uparrow \varphi \qquad \varphi \uparrow \qquad \uparrow \varphi$$
$$F + F \to F \qquad F \underset{\lambda}{\to} F$$

Example: (v) Let A be a closed set in X. Define $F_A = F/_{F_{X-A}}$.

Clearly $(F_A)_x = \begin{cases} F_x & \text{if } x \in A \\ 0 & \text{if } x \notin A \end{cases}$

For any sheaf map $\varphi: F \to G$ we can define sheaves which we call Imφ, Ker φ, Coim φ, and Coker φ (or Ckr φ). We define $(\text{Ker } \varphi)_x = \varphi^{-1}(0_x)$. The set of zero elements of G is open in G, therefore Ker φ is open in F, and by proposition 1, Ker φ is a subsheaf of F. Coim φ is by definition $F/_{\text{Ker } \varphi}$.

φ is an open map. Therefore Imφ is a subsheaf of G. Coker φ is by definition, $G/_{\text{Im}\varphi}$.

Direct sums of sheaves exist; cf. §4.1 below. Clearly both the categories of sheaves and of protosheaves are exact categories. We shall say that a sequence $F \xrightarrow{\varphi} G \xrightarrow{\psi} H$ is exact (at G) if Imφ = Ker ψ. A sequence

$$\to G_n \xrightarrow{\varphi_n} G_{n+1} \xrightarrow{\varphi_{n+1}} G_{n+2} \to \cdots$$

of a finite or infinite number of sheaves is exact if it is exact at each G_n. In the case of a finite sequence,

$$G_0 \xrightarrow{\varphi_0} G_1 \to \cdots \xrightarrow{\varphi_n} G_n ,$$

this implies nothing about Ker φ_0 or Im φ_n.

§2

§2.1 Sections. Let (F,X,p) be a sheaf and U an open set in X. A section over U is a continuous map $s: U \to F$ such that ps = identity: $U \to U$. We define the zero section over U to be the function $0: U \to F$ given by $0(x) = 0_x \in F_x$. Recall $\Gamma(U,F)$ is the set of sections $U \to F$.

For any $y \in F$, there exists a section over some $V \subset X$ passing through y. Take V to be a homeomorph under p of some neighborhood W of y and let $s = (p \mid W)^{-1}$.

Lemma 1: Let $s,t \in \Gamma(U,F)$. Then the set $A = \{x \in X; s(x) = t(x)\}$ is open, or equivalently if two sections agree at x they agree in a neighborhood of x.

Proof: Let W be a neighborhood of s(x) which is mapped homeomorphically by p onto a neighborhood of x. There exists a neighborhood V of x such that $s(V), t(V) \subset W$. If $y \in V$, then $s(y) = t(y)$ because there is only one point of W lying over y.

Remarks: If F is Hausdorff and two sections over a connected, open set $U \subset X$ agree at a point of U, then they agree over all of U. For given two maps into a Hausdorff space, the set of points on which they agree is closed. If the base is Hausdorff and locally connected, the converse is true. (We leave the proof to the reader.)

If s,t are 2 sections over U, define s+t by $(s+t)(x) = s(x) + t(x)$, $(ks)(x) = ks(x)$. Thus the set $\Gamma(U,F)$ of sections of F over U is a K-module. For a fixed U, $\Gamma(U, \quad)$ is a covariant K-functor from K-sheaves to K-modules.

For if $\varphi : F \to G$ is a map and $s: U \to F$ is a section of F over U, then $\varphi s: U \to G$ is a section of G over U, giving a map

$$\Gamma(U,\varphi) : \Gamma(U,F) \to \Gamma(U,G).$$

If open $V \subset U$, define $\rho_V^U : \Gamma(U,F) \to \Gamma(V,F)$ by restriction; i.e., if we have section $s: U \to F$ $\rho_V^U(s) = s \mid V: V \to F$. If $\varphi: F \to G$ and $V \subset U$, the following diagram is commutative.

$$\begin{array}{ccc} \Gamma(U,F) & \to & \Gamma(U,G) \\ \downarrow & & \downarrow \\ \Gamma(V,F) & \to & \Gamma(V,G) \end{array}$$

$\Gamma(\quad ,F)$ is a contravariant functor from open sets and inclusion maps to K-modules, for clearly ρ_U^U = identity, and if

$$W \subset V \subset U, \text{ then } \rho_W^U = \rho_W^V \rho_V^U$$

§2.2 Stacks (older name, pre-sheaf).

More generally, a stack is a contravariant functor

from open sets and inclusion maps (of some fixed space X) to the category of K-modules.

A map of stacks is a natural transformation of functors. These maps $\underline{S} \to \underline{T}$ form a K-module by $(\alpha f+\beta g)(U) = \alpha f(U)+\beta g(U)$ for $\alpha,\beta \in K$ and $f,g: \underline{S} \to \underline{T}$.

The stacks with these maps form a K-category, which is also exact, and the functor $\Gamma(\ \ ,x)$ which takes F to $\Gamma(\ \ ,F)$ is a K-functor from sheaves to stacks.

§ 2.21

Given a stack $\underline{S}$, for each $x \in X$, define $S_x = \varinjlim_{U \in x} \underline{S}\,(U)$ and $\varphi_x^U : \underline{S}\,(U) \to S_x$ to be the natural map. (See chapt. I).

Define a protosheaf $\overline{S} = \bigcup_{x \in X} S_{x}$ with projection p. For each $s \in \underline{S}\,(U)$ define a section $\overline{s}$ of $\overline{S}$ by $\overline{s}\,(x) = \varphi_x^U(s)$. Form a sheaf S from $\overline{S}$ by taking as a base for neighborhoods in $\overline{S}$ the set $\{\overline{s}(U);$ all U, all $s \in \underline{S}(U)\}$. p is onto; we show it is continuous and a local homeomorphism.

Let U open $\subset X$, $y \in p^{-1}(U)$, and $p(y) = x$. Then from the construction of the direct limit, there exists $V \subset U$, $s \in \underline{S}\,(V)$ such that $\varphi_x^V(s) = y$. Then $y \in \overline{s}\,(V)$, since $\overline{s}(x) = \varphi_x^V(s) = y$. Therefore $p^{-1}(U)$ is open. If $p' = p \mid \overline{s}(V)$, $p'\overline{s}$ = identity and $\overline{s}\,p'$ = identity. Therefore p' is a homeomorphism since $\overline{s}$ is continuous.

We leave to the reader the proof that addition and multiplication by elements of K are continuous.

Note that we could define stacks and sheaves of sets; i.e., no algebraic structure is assumed on the stalks, and we can still derive as above a construction of a sheaf of sets from a stack of sets. Or we may repeat the construction for any algebraic structure which admits direct limits.

Given any map $\underline{\eta} : \underline{S} \to \underline{T}$ of stacks, we obtain, by passage to direct limits (which we know is a functor) maps $\eta_x: S_x \to T_x$ for each $x \in X$. These can be stuck together to give a map $\overline{\eta} : \overline{S} \to \overline{T}$ of protosheaves. This map is in fact consistent with the topology given to $\overline{S}$ and $\overline{T}$; let $\eta(x) = y$, and $\overline{t}(U)$ any basic neighborhood of y. Let $V \subset U$ be a neighborhood of p(x) so small that x has a preimage s in $\underline{S}(V)$.

Then $\underline{\eta}$ (V) s and $\rho_V^U t$ are sections of T passing through y. Therefore they agree on a neighborhood W of p(y). So ρ_W^V s (W) is a neighborhood of x which is mapped into $\overline{t}(U)$ by η.

So we have a map η: S → T of sheaves, or, to summarize, there exists a covariant functor L from stacks to sheaves.

§2.23

What is the relation of L and Γ (,*)? We have the proposition:

Proposition 3: LΓ(,*) is naturally equivalent to the identity. There exists a natural transformation: identity → Γ(, *) L.

The latter natural transformation is $\underline{S}$ (U) → Γ(U,S) where $\underline{s} \to s$ by $s(x) = \varphi_x^U(\underline{s})$.

For any sheaf F, we define maps $F \underset{j}{\overset{i}{\leftrightarrows}} L\Gamma(,F) = F'$.

Let y ∈ F, s any section through y; i.e., s ∈ Γ(U,F) for some U.

Define $i(y) = \varphi_{p(y)}^U(s) \in F'_{p(y)}$.

Let z ∈ F'. Take some w ∈ Γ(U,F) such that $\varphi_x^U(w) = z$.

Define j(z) = w(x).

We leave the reader to check that these give well defined natural transformations of functors and that ij and ji are identity maps.

Proposition 4: L is an exact functor; Γ(,*) is left-exact. We leave the proof to the reader.

Many of the most interesting sheaves, such as those of use in algebraic geometry, can be defined most naturally via stacks.

Example:(vi) Sheaf of germs of functions

Let M be a K-module; we define a stack $\underline{S}$ by letting $\underline{S}$(U) be the set of functions U → M and give $\underline{S}$(U) the natural K-module structure. If V ⊂ U, define φ_V^U : S(U) → S(V) by restriction. The associated sheaf S is the sheaf of germs of functions with values in M. That is to say, let x ∈ X, U, V neighborhoods of x, and f,g functions f: U → M, g:V → M. We say that f and g are locally equal if there is a neighborhood W of x, W ⊂ U ∩ V such that f | W = g | W.

This is an equivalence relation; the equivalence class of f is called the germ of f at x. The set of germs at x is the stalk S_x. In this case $\Gamma(\ \ ,S) = \underline{S}$.

For the proof see Chapter VI.

(vii) We can generalize the above as follows. If M is any K-module, U open in X:

Let $\Psi^p(U,M)$ be the set of functions $U^{p+1} \to M$. If $V \subset U$, define $\Psi^p(U,M) \to \Psi^p(V,M)$ by restriction. $\Psi^p(U,M)$ has a natural K-module structure, and example (vi) is given by p = o. The associated sheaf of $\Psi^p(\ \ ,M)$ we write C^p_M, the Alexander-Spanier sheaf of p-dimensional cochains.

Define a map the "coboundary" $\delta(U) : \Psi^p(U,M) \to \Psi^{p+1}(U,M)$ as follows: If $f\colon U^{p+1} \to M$, let

$$(\delta(U)f)(x_o,\ldots,x_{p+1}) = \sum_{o}^{p+1} (-1)^i f(x_o,\ldots, \hat{x}_i,\ldots, x_{p+1}).$$

δ is a map of stacks, and $\delta\delta = 0$.

If $\Psi^{-1}(\ \ ,M)$ is the constant stack, define $\epsilon(U)\colon \Psi^{-1}(U,M) \to \Psi^o(U,M)$ by letting $\epsilon(U)(m)$ be the constant map with value m. This is called the augmentation. Note that $\delta\,\epsilon = 0$. So we get what is known as an augmented cochain complex of stacks, i.e., a sequence

$$M \xrightarrow{\epsilon} \Psi^o(\ \ ,M) \xrightarrow{\delta} \Psi^1(\ \ ,M) \xrightarrow{\delta} \Psi^2(\ \ ,M) \to \cdots$$

with $\delta^2 = 0$, $\delta\epsilon = 0$.

This sequence is in fact exact. We define maps $s\colon \Psi^{i+1}(U,M) \to \Psi^i(U,M)$ $(i \geq 0)$ and $\tau\colon \Psi^o(U,M) \to \Psi^{-1}(U,M)$. Let $a \in U$. Define

$$(sf)(x_o,\ \ldots,\ x_i) = f(a,x_o,\ \ldots,\ x_i),\quad \tau f = f(a).$$

The reader will easily prove that $s\delta + \delta s = 1$ except in $\psi^o(U,M)$, where $s\delta = 1 - \epsilon\tau$.

By applying L we get an augmented, acyclic cochain complex of sheaves.

$$M \xrightarrow{\epsilon} C^o_M \xrightarrow{\delta} C^1_M \xrightarrow{\delta} C^2_M \to \cdots$$

(viii) <u>Sheaf of singular cochains</u>

Define $C^n(U,M)$ to be the module of singular n-dimensional cochains of U. We have the sequence, where δ and ϵ are the usual coboundary and augmentation,

$$M \xrightarrow{\epsilon} C^{o}(U,M) \xrightarrow{\delta} C^{1}(U,M) \xrightarrow{\delta} C^{2}(U,M) \rightarrow \cdots$$

giving the sheaf of singular cochains

$$M \xrightarrow{\epsilon} C^{o}_{M} \xrightarrow{\delta} C^{1}_{M} \xrightarrow{\delta} C^{2}_{M} \rightarrow \cdots$$

This sequence is not exact in general. If it is we say the space is HLC.

(ix) Sheaf of singular chains

Let $C_n(X, X-U; M)$ be the module of n-dimensional singular chains of (X,X-U) with coefficients in M. This is a stack; we have the sequence

$$M \xleftarrow{\epsilon} C_o(X,X-U;M) \xleftarrow{\partial} C_1(X,X-U;M) \xleftarrow{\partial} C_2(X,X-U;M) \leftarrow \cdots$$

which gives the sheaf of singular chains.

We can also define this using locally finite singular chains. For this definition and further properties of the sheaf, see Chapter VI.

§ 3

§ 3.1 Restriction. Let $A \subset X$ be any subset of X. Let F be a sheaf over X. Define $F \mid A = p^{-1}(A)$. Then $F \mid A$ is a sheaf over A, the projection, topology, and algebraic structure being induced by that of F. We call $F \mid A$ the restriction of F to A.

§ 3.2 Prolongation by Zero

Let $A \subset X$ be any subset of X and F' be a sheaf over A. We say that a sheaf F over X is a prolongation of F' by zero if $F \mid A = F'$ and $F \mid (X-A) = 0$.

Proposition 5: There is, up to natural isomorphism, at most, one prolongation of F' by zero. Such a prolongation always exists if A is locally closed in X, where

Definition: A is locally closed in X if each $x \in A$ has a neighborhood N_x such that $N_x \cap A$ is closed in N_x.

Proof: Clearly the protosheaf $\overline{F}$ is unique and always exists. Suppose we have two topologies $\mathcal{J}_1$ and $\mathcal{J}_2$ on $\overline{F}$ which make it a sheaf with the required properties. We show $\mathcal{J}_1 =$

$\mathcal{J}_2$. Let W_1', W_2' be sets containing $y \in \overline{F}_x$ and open in the respective topologies. We find a set open in both topologies containing y and contained in both W_1' and W_2'. Choose a neighborhood W_1 of y so that $W_1 \subset W_1'$ and $p \mid W_1$ is a homeomorphism onto a neighborhood U_1 of x. Similarly, choose a $W_2 \subset W_2'$ which is homeomorphic in its topology to a neighborhood U_2 of x. Define sections $s_1 \in \Gamma(U_1,\overline{F})$ and $s_2 \in \Gamma(U_2,\overline{F})$ as the inverses of the homeomorphisms. Now $s_1 \mid A$ and $s_2 \mid A$ are continuous sections of F', so if $x \in A$, there exists a set V' open in A containing x and such that $s_1 \mid V' = s_2 \mid V'$. Then $V' = V \cap A$ where V is open in X. If $U = U_1 \cap U_2 \cap V$, then $s_1 \mid U = s_2 \mid U$, since $\overline{F} \mid X\text{-}A = 0$. Also U is not empty, since $x \in U$ and is open in X. Then $s_1(U) = s_2(U)$ is contained in both W_1' and W_2', and is open in both topologies, since s_1 and s_2 are open maps in their respective topologies. If $x \notin A$, then y is the zero over x. The zero section over a sufficiently small neighborhood of x will be open in both topologies and contained in both W_1' and W_2'. Therefore $\mathcal{J}_1 = \mathcal{J}_2$.

We now assume A is locally closed and define a topology on $\overline{F}$. Let $y \in \overline{F}_x$. If $x \notin A$, we define the basic neighborhoods of y to be the zero sections over heighborhoods of x. If $x \in A$, there is a heighborhood N_x of x such that there is a section $s \in \Gamma\,(A \cap N_x,\ F)$ with $s(x) = y$. We define a basic neighborhood (N_x,s) of y to be the union of $s(A \cap N_x)$ with the zero section over $N_x - A$. Clearly the intersection of two such basic neighborhoods of y contains a third, for two sections of F through y agree on small enough neighborhoods of x. To show that these neighborhoods define a topology on $\overline{F}$, we must show that if V is a neighborhood of y it is also a neighborhood of all points in some smaller neighborhood W of y. This is trivial if V is a zero section. Let $V = (N_x,s)$ with $x \in A$. Since A is locally closed, there is an open neighborhood M_x of x such that $M_x \subset N_x$ and $M_x \cap A$ is closed in M_x. Define $W = (M_x,\ s \mid M_x \cap A)$. If $z \in W$ and $p(z) \in A$, then W is also a neighborhood of z. If $z \in W$ and $p(z) \notin A$, then p(z) has a neighborhood U disjoint from

from $M_x \cap A$. Therefore, $p^{-1}(U) \cap W$ is the zero section over U and so W is a neighborhood of z.

Definition: If A is locally closed in X and F a sheaf over A, define F^X to be this unique prolongation by zero of F.

Definition: If A is locally closed in X, F a sheaf over X, define $F_A = (F \mid A)^X$. Note that this is consistent with our earlier definitions. For A is locally closed is equivalent to A being the intersection of an open and a closed set of X. Thus all open, and all closed sets are locally closed.

Remark: The above operations we can write functorially as $*_A$, $* \mid A$, $*^X$. They are all exact functors from sheaves to sheaves; in the last two cases the domain and range are sheaves over different spaces.

§4 Various universal constructions

§4.1 Direct sums

In any K-category a direct sum H of F and G is characterized by maps

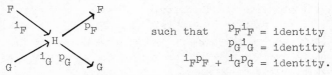

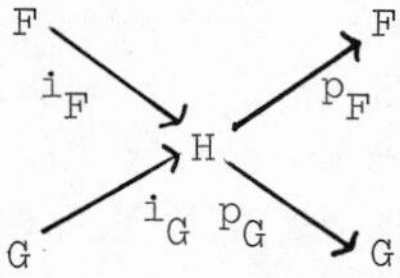

such that

$$p_F i_F = \text{identity}$$
$$p_G i_G = \text{identity}$$
$$i_F p_F + i_G p_G = \text{identity}.$$

It follows that direct sums are preserved by K-functors. Now let $\underline{S}$ and $\underline{T}$ be stacks and let $\underline{S} + \underline{T}$ be defined by $(\underline{S} + \underline{T})(U) = \underline{S}(U) + \underline{T}(U)$ (direct sum). This is easily shown to be a direct sum of stacks. Since L is a K-functor, $L(\underline{S} + \underline{T}) = S + T$, where $S + T$ is a direct sum of sheaves. A direct definition of this is the one given earlier, viz. $S + T = \{(y,z) \in S \times T;\ p(y) = p(z)\}$. These definitions are, by the results of Chapter I, equivalent.

§4.2 Tensor products

Before we can define this, we must say what we mean by bilear maps.

a. Let F,G,H be sheaves. A bilinear map

$$f: F + G \to H$$

shall (1) be continuous,

(2) make the diagram

$$\begin{array}{ccc} F + G & \rightarrow & H \\ & \searrow \; X \; \swarrow & \end{array}$$

commutative, and

(3) be such that $f \mid F_x + G_x$ is a bilinear map of K-modules.

b. Let $\underline{F}$, $\underline{G}$, $\underline{H}$ be stacks. A bilinear map

$$\underline{f}\colon \underline{F} + \underline{G} \rightarrow \underline{H}$$

is a collection of bilinear maps $\underline{f}_U\colon \underline{F}(U) + \underline{G}(U) \rightarrow \underline{H}(U)$ such that if $V \subset U$, the following diagram is commutative:

$$\begin{array}{ccc} \underline{F}(U) + \underline{G}(U) & \rightarrow & \underline{H}(U) \\ \downarrow & & \downarrow \\ \underline{F}(V) + \underline{G}(V) & \rightarrow & \underline{H}(V) \end{array}$$

Since direct limits preserve direct sums and bilinearity, $\underline{f}$ gives rise to a corresponding bilinear map $\overline{f}\colon \overline{F} + \overline{G} \rightarrow \overline{H}$ of protosheaves and an argument exactly the same as before shows that f is consistent with the topology given to $\overline{F}$, $\overline{G}$, and $\overline{H}$. So we have a bilinear map of sheaves $f\colon F + G \rightarrow H$; i.e., L preserves bilinearity.

So, in fact, does $\Gamma(\ \ ,*)$. For if $f\colon F + G \rightarrow H$ is a bilinear map of sheaves, $\Gamma(\ \ , F + G) = \Gamma(\ \ ,F) + \Gamma(\ \ ,G)$ since $\Gamma(\ \ ,*)$ is a K-functor and composition with f gives a bilinear map

$$\Gamma(\ \ ,F) + \Gamma(\ \ ,G) \rightarrow \Gamma(\ \ ,H).$$

<u>Proposition 6</u>: If $\underline{F}$, $\underline{G}$ are stacks, there exists a universal bilinear map $\underline{i}\colon \underline{F} + \underline{G} \rightarrow \underline{H}$.

<u>Definition</u>: $\underline{F} \otimes \underline{G} = \underline{H}$.

To prove the proposition, let $\underline{H}(U) = \underline{F}(U) \otimes_K \underline{G}(U)$ and let the map $\underline{i}(U)\colon \underline{F}(U) + \underline{G}(U)$ be the usual one. Then if $\underline{F} + \underline{G} \xrightarrow{f} \underline{T}$ is bilinear, we have for each $U^{open} \subset X$ a unique factorization $\underline{H}(U) \xrightarrow{h(U)} \underline{T}(U)$.

By considering the diagram,

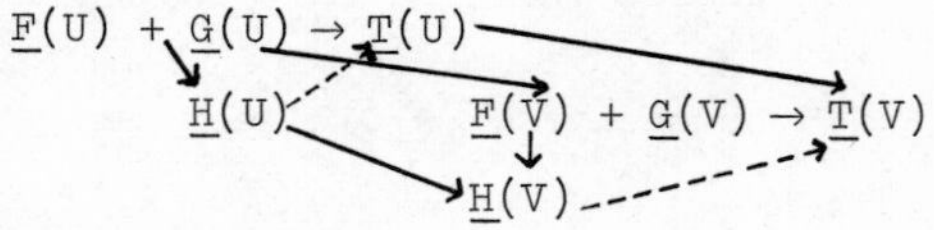

where $V \subset U$, we easily see that $\underline{h}\colon \underline{H} \rightarrow \underline{T}$ is a natural transformation, i.e., is a map of stacks.

Proposition 7: If F, G are sheaves, there exists a universal bilinear map i: $F + G \to H$.

Definition: $F \otimes G = H$.

There exist stacks $\underline{F}$, $\underline{G}$ such that $L(\underline{F}) = F$, $L(\underline{G}) = G$. (e.g., $\underline{F} = \Gamma(\ ,F)$).

Let $H = L(\underline{F} \otimes \underline{G})$.

Let f: $F + G \to H'$ be bilinear.

Consider the following diagram:

$$\begin{array}{ccc} \underline{F} + \underline{G} \to \Gamma(\ ,F) + \Gamma(\ ,G) & \xrightarrow{\Gamma(\ ,f)} & \Gamma(\ ,H') \\ \underline{i} \searrow & & \nearrow \underline{h} \\ & \underline{F} \otimes \underline{G} & \end{array}$$

α is the natural transformation given in proposition 3. Since $\underline{i}$ is universal, this diagram has a unique completion $\underline{h}$ as shown. Applying L, we obtain a sheaf map

$$h: H \to H' \text{ such that } f = h \circ i.$$

Direct limits commute with tensor products. Therefore $H_x = F_x \otimes_K G_x$. But the factorization $F_x + G_x \to H'_x$ is unique.

$$\begin{array}{ccc} F_x + G_x & \to & H'_x \\ & \searrow & \nearrow \\ & H_x & \end{array}$$

Therefore h is unique.

Proposition 8: $\otimes$ is right exact on the category of stacks; and so, by application of L, $\otimes$ is right exact on the category of sheaves.

The proof is easy.

§ 5. Supports

Let F be a sheaf, $s \in \Gamma(U,F)$. We define the support of s, written $|s|$, to be the set $\{x \in U;\ s(x) \neq 0\}$. This is a closed set in U. For if $s(x) = 0$, since p is a local homeomorphism, there is a small neighborhood of x mapped entirely to zero by s; i.e., $U - |s|$ is open.

Let $f: F \to G$ be a sheaf map. We define $|f|$, support of f, to be the closure of $\{x \in X;\ f \mid F_x \neq 0\}$.

For protosheaves, we modify the above by saying that the support of a section shall be the closure of the set of x not mapped into 0.

Lemma 2: Let f: $F \to G$ be a map that $|f| \overset{\text{closed}}{\subset} C \subset X$. Then there is a unique factorization

$$\begin{array}{ccc} F & \xrightarrow{f} & G \\ \downarrow & \swarrow & \\ F_C & & \end{array}$$

Note that F_C is a quotient sheaf of F; i.e., $F \to F_C$ is a universal map for maps $F \to G$ with support contained in C.

For $F \to F_C$ is epi, and f annihilates the kernel of this this.

§ 5.2 Family of supports

A family Φ of subsets of X is called a family of supports if:

Axiom (1) $A \in \Phi \Rightarrow A$ closed.

Axiom (2) $B \overset{\text{closed}}{\subset} A$ and $A \in \Phi \Rightarrow B \in \Phi$.

Axiom (3) $A, B \in \Phi \Rightarrow A \cup B \in \Phi$.

Note: Cartan used two additional restrictions in Φ.

Φ is called paracompactifying (abbreviated here to PF) if in addition:

Axiom (4) $A \in \Phi \Rightarrow A$ paracompact.

Axiom (5) $A \in \Phi \Rightarrow A$ has a neighborhood in Φ.

A paracompact space is an Hausdorff space in which every covering has a locally finite refinement. By "A has a neighborhood in Φ" we mean there is a $B \in \Phi$ such that $A \subset \operatorname{Int} B$.

Definition: If F is a sheaf and Φ a family of supports, define $\Gamma_\Phi(F) = \{s \in \Gamma(X,F);\ |s| \in \Phi\}$. $\Gamma_\Phi(F)$ is the set of sections with support in Φ, and it is easily seen to be a submodule of $\Gamma(X,F)$. Let $f: F \to G$. Then $|fs| \subset |s|$. So we have a map $\Gamma_\Phi(f): \Gamma_\Phi(F) \to \Gamma_\Phi(G)$. Clearly $\Gamma_\Phi(\ \ )$ is a convariant K-functor from sheaves to modules.

Definition: If F, G are sheaves, define $\operatorname{Hom}_\Phi(F,G)$ to be the set of sheaf maps $F \to G$ with support in Φ. $\operatorname{Hom}_\Phi(\ \ ,\ \ )$ is a K-functor of two variables from sheaves to modules, covariant in one variable, contravariant in the other.

Proposition 9: $\operatorname{Hom}_\Phi(\ \ ,\ \ )$ is left-exact in either variable. Let $0 \to F' \xrightarrow{i} F \xrightarrow{j} F'' \to 0$ be an exact sequence of sheaves. Let G be a sheaf.

Consider the induced sequence,

$$0 \to \mathrm{Hom}_\Phi(G,F') \xrightarrow{i'} \mathrm{Hom}_\Phi(G,F) \xrightarrow{j'} \mathrm{Hom}_\Phi(G,F'').$$

Let $s' \in \mathrm{Hom}_\Phi(G,F')$. i' is mono.
For $i's' = 0 \Rightarrow (i(s'(y)) = 0$ all $y \in G$
$\Rightarrow s'(y) = 0 \Rightarrow s' = 0$ since i is mono.
Clearly, $(j'i'(s'))\,(y) = (ji)(s'(y)) = 0$ since $ji = 0$.
Let $s \in \mathrm{Hom}_\Phi(G,F)$ and $j's = 0$. Then $j'(s(y)) = 0$, $y \in G$.
Since i is mono and the first sequence is exact, there exists a unique $s'(y) \in F'$ such that $i(s'(y)) = s(y)$.

The reader can check that s' is a sheaf map. $|s'| = |s| \to \in \Phi$ since i is mono. Therefore $s' \in \mathrm{Hom}_\Phi(G,F')$ with $i'(s') = s$.
Therefore the sequence above is exact.
In a similar way, we prove $\mathrm{Hom}_\Phi(\ ,\ )$ is left-exact in the other variable.
Corollary: $\Gamma_\Phi(\ )$ is left-exact.
For $\Gamma_\Phi(F) = \mathrm{Hom}_\Phi(K,F)$ naturally.

§ 6

Finally, we define a functor from protosheaves to sheaves. If $\overline{M}$ is a protosheaf, $\Gamma(\ ,\overline{M})$ is a stack. We write $\tilde{\overline{M}} = L\Gamma(\ ,\overline{M})$.
Lemma 3: If F is a sheaf and $\overline{M}$ a protosheaf, then there is a natural support preserving isomorphism

$$\Gamma\colon \mathrm{Hom}\,(\overline{F},\overline{M}) \approx \mathrm{Hom}\,(F, \tilde{\overline{M}});$$

i.e., if $f\colon \overline{F} \to \overline{M}$ is a map of protosheaves, $|f| = |\Gamma(f)|$.

This shows incidentally that ~ is a functor. We express the lemma by saying that - and ~ are adjoint functors. Another example of adjoint functors are Ω and E as functors of topological spaces; for map $(X, \Omega Y)$ = map (EX,Y). For more details on adjoint functors see Kan (3).
Proof of Lemma: We construct K-homomorphisms.

$$\mathrm{Hom}(\overline{F},\overline{M}) \underset{\xi}{\overset{\eta}{\rightleftarrows}} \mathrm{Hom}(F,\tilde{\overline{M}}).$$

i. Let $f \in \mathrm{Hom}(F,M)$. The composition (for each $U \subset X$)

$$\Gamma(U,F) \xrightarrow{i} \Gamma(U,\overline{F}) \xrightarrow{\Gamma(U,\overline{f})} \Gamma(U,\overline{M})$$

where i is the inclusion map gives by application of L, a map

$$g = \eta(\overline{f}) : F \to \tilde{\tilde{M}}.$$

Explicitely, let $\alpha \in F_x$ and s be a local section through α.

Then $g(\alpha) = \varphi_x^U$ (fs) for some neighborhood U of x.

Suppose $x \notin |\overline{f}|$, then $\overline{f} \mid F_y = 0$ for any y in a neighborhood of x.

Therefore $\overline{f}s = 0$ in a neighborhood of x.

Therefore $\varphi_x^U (\overline{f}s) = 0$; i.e., $g(\alpha) = 0$ and $g \mid F_x = 0$.

Therefore $|g| \subset |\overline{f}|$.

ii. Any $x \in U$ defines a map $\Gamma(U,\overline{M}) \to \overline{M}_x$ by $s \to s(x)$. By passage to direct limits, we have a map $\tilde{\overline{M}}_x \to \overline{M}_x$ and so a map of protosheaves $\chi : \tilde{\tilde{M}} \to \overline{M}$.

If $g \in \mathrm{Hom}(F,\tilde{\tilde{M}})$, define $\xi(g) = \chi\,\overline{g}$. $\overline{F} \xrightarrow{\overline{g}} \tilde{\tilde{M}} \xrightarrow{\chi} \overline{M}$.

I say $|\chi\overline{g}| \subset |g|$.

For if $x \notin |g|$, and $\alpha \in \overline{F}_x$, then $\chi\overline{g}(\alpha) = \chi(0) = 0$.

Explicitely, let $\alpha \in \overline{F}_x$ and let t_α be a section of $\overline{M}$ so that $\varphi_x^U(t_\alpha) = g(\alpha)$ for some neighborhood U of x.

Then $\chi\overline{g}.(\alpha) = t_\alpha(x)$.

iii. We show $\xi\eta$ = identity and $\eta\xi$ = identity.

Let $\eta(f) = g$, $\xi(g) = h$. Let α, t_α be as in (ii).

Then $h(\alpha) = t_\alpha(x) = (fs)(x)$ (for some section s, as in (i)) $= f(s(x)) = f(\alpha)$.

Therefore h = f and $\xi\eta$ = identity.

Let $\xi(g) = f$ and $\eta(f) = h$. Let $\alpha \in F_x$. Let s be a local section of F through α. Then fs is a local section of $\overline{M}$ and $h(x) = \varphi_x^U$ (fs) for some neighborhood U of x.

But we can choose s so that $\varphi_x^U (fs) = g(\alpha)$.

Thus $h(\alpha) = g(\alpha)$.

The naturality of η follows from its functorial definition

§ 7

Definition: We say a sheaf (or module) I is injective if,

for any exact sequence $0 \to A \xrightarrow{i} B$ and map $f{:}A \to I$, there is a map $h{:}B \to I$ such that $hi = f$.

$$\begin{array}{ccccc} 0 & \to & A & \xrightarrow{i} & B \\ & & f\downarrow & \swarrow h & \\ & & I & & \end{array}$$

Lemma 4: There is always a completion h with $|h| \subset |f|$.

Proof: The diagram
$$\begin{array}{ccccc} 0 & \to & A & \to & B \\ & & f\downarrow & & \\ & & I & & \end{array}$$
has by lemma 2 with $C = |f|$ a factorization, given by the diagram

$$\begin{array}{ccccc} 0 & \to & A & \to & B \\ & & & & \downarrow \\ 0 & \to & A_C & \to & B_C \\ & \swarrow & & \swarrow & \\ I & & & & \end{array}$$

Since I is injective, there is a completion $h'{:}B_C \to I$. If h is the composition $B \to B_C \xrightarrow{h'} I$, $|h| \subset C$.

Theorem 1: Any module can be imbedded in an injective module. For the proof we refer the reader to Cartan and Eilenberg, p. 9 or 51, or Eckmann-Schöpf (4).

Theorem 2: Every sheaf can be imbedded in an injective sheaf.

Proof: Let F be a sheaf. A protosheaf is injective if and only if its stalks are injective.

Lemma 5: Protosheaf $\bar{I}$ is injective $\Rightarrow \tilde{\bar{I}}$ is injective.

For let the row of
$$\begin{array}{ccccc} 0 & \to & A & \xrightarrow{g} & B \\ & & f\downarrow & & \\ & & \tilde{\bar{I}} & & \end{array}$$
be exact.

Now the diagram
$$\begin{array}{ccccc} 0 & \to & \bar{A} & \xrightarrow{\bar{g}} & \bar{B} \\ & & f'\downarrow & & \\ & & \bar{I} & & \end{array}$$
certainly has a completion $\bar{h}{:}\ \bar{B} \to \bar{I}$, since $\bar{I}$ is injective.

By lemma 3 there is a completion of the first diagram by $h{:}\ B \to \tilde{\bar{I}}$. For this theorem, choose $\bar{I}$ to be a protosheaf of injective modules such that $I_x \supset F_x$. The map $\bar{F} \xrightarrow{i} \bar{I}$ gives, by lemma 3, a map $F \to \tilde{\bar{I}}$. Let $G = \operatorname{Ker}(F \to \tilde{\bar{I}})$. Then the composition $\bar{G} \to \bar{F} \xrightarrow{i} \bar{I}$ is zero. But I is mono. Therefore $\bar{G} = 0$. Therefore $G = 0$.

We could define a projective sheaf P to be one such that for any exact sequence $A \to B \to 0$ and any map $f: P \to B$, there is a map $h: P \to A$ making the diagram

$$\begin{array}{ccccc} & {}^{h}\swarrow & P & {}^{f} & \\ A & \rightarrow & B & \rightarrow & 0 \end{array}$$

commutative. But if the base space X is not discrete, I know of no examples of projective sheaves except the zero sheaf.

III. HOMOLOGICAL ALGEBRA

To construct a homology theory for sheaves and to prove its uniqueness we need a number of concepts and propositions from homological algebra; we assume a certain familiarity with the techniques involved. We consider two exact K-categories $\mathfrak{A}$ and $\mathfrak{B}$ (e.g., sheaves, modules). Functors will always, unless stated otherwise, be K-functors $\mathfrak{A} \to \mathfrak{B}$. We refer to Cartan and Eilenberg's "Homological Algebra" as CE.

§ 1

A covariant δ-functor T^* is a series of covariant functors T^i, $a < i < b$, and for each exact sequence $0 \to A' \to A \to A'' \to 0$ a map

$$\delta: \ T^i(A'') \to T^{i+1}(A') \qquad a < i < b-1$$

such that (i) δ is natural. By this we mean the following:

A map of exact sequences $E \to E'$ is a commutative diagram

$$\begin{array}{ccccccccc} 0 & \to & A' & \to & A & \to & A'' & \to & 0 \\ & & \downarrow & & \downarrow & & \downarrow & & \\ 0 & \to & B' & \to & B & \to & B'' & \to & 0 \end{array}$$

where E and E' are the upper and lower exact sequences of this diagram.

Condition (i) states that the induced diagram

$$\begin{array}{ccc} T^i(A'') & \xrightarrow{\delta} & T^{i+1}(A') \\ \downarrow & & \downarrow \\ T^i(B'') & \xrightarrow[\delta]{} & T^{i+1}(B') \end{array}$$

shall be commutative.

(ii) From an exact sequence $0 \to A' \to A \to A'' \to 0$ we obtain a sequence

$$\cdots \xrightarrow{\delta} T^i(A') \to T^i(A) \to T^i(A'') \xrightarrow{\delta} T^{i+1}(A') \to \cdots$$

We require that the composition of any two adjacent maps in this sequence shall be zero.

<u>Definition</u>: A contravariant δ-functor T^* is a series of contravariant functors T^i, $a < i < b$, and for each exact sequence $0 \to A' \to A \to A'' \to 0$ a natural map $\delta\colon T^i(A') \to T^{i+1}(A'')$ such that in the sequence

$$\ldots \to T^i(A'') \to T^i(A) \to T^i(A') \xrightarrow{\delta} T^{i+1}(A'') \to \ldots$$

The composition of two adjacent maps is zero.

Assume from now on that δ-functors are covariant.

<u>Definition</u>: A map of δ-functors $(T^i)_{a < i < b}$ and $(S^i)_{a < i < b}$ is a set of natural transformations $f^i\colon T^i \to S^i$ which preserve δ; i.e., if $0 \to A' \to A \to A'' \to 0$ is exact, the following diagram is commutative:

$$\begin{array}{ccccc} & T^i(A'') & \xrightarrow{\delta} & T^{i+1}(A') & \\ f^i(A'') & \downarrow & & \downarrow & f^{i+1}(A') \\ & S^i(A'') & \xrightarrow{\delta} & S^{i+1}(A') & \end{array}$$

<u>Definition</u>: A δ-functor T^* is <u>exact</u> if the sequence in (ii) is exact.

<u>Definition</u>: Let S be a (covariant) K-functor. An augmented δ-functor T^* over S is a δ-functor $(T^i)_{o \leq i < a}$ together with a natural transformation $\epsilon\colon S \to T^o$.

A map of augmented δ-functors $\{U^i\}_{o \leq i < a}$ over R, $\{V^i\}_{o \leq i < a}$ over S, over a natural transformation $\tau\colon R \to S$ is a map f of δ-functors such that the following diagram commutes

$$\begin{array}{ccc} R & \xrightarrow{\epsilon} & U^o \\ \tau \downarrow & & \downarrow f^o \\ S & \xrightarrow{\epsilon} & V^o \end{array}$$

<u>Definition</u>: A δ-functor $T^* = (T^i)_{o \leq i < a}$ is universal if for any δ-functor $S^* = (S^i)_{o \leq i < a}$ and any natural transformation $f^o : T^o \to S^o$ there exists a unique map $g\colon T^* \to S^*$ of δ-functors such that $g^o = f^o$.

<u>Definition</u>: A functor F is <u>effaceable</u> if for each $A \in \mathcal{A}$, there is a monomorphism $i\colon A \to M$ such that $F(i) = 0$.

A monomorphism $i\colon A \to M$ with $F(i) = 0$ is called an <u>effacement</u> of F at A.

<u>Definition</u>: A cohomological δ-functor T^* over S is an

augmented δ-functor over S which is exact, and such that T^i is effaceable for $i > o$ and $\epsilon: S \to T^o$ is a natural equivalence.

<u>Theorem 1</u>: If $T^* = (T^i)_{o \leq i < a}$ is an exact, δ-functor with T^i effaceable for $i > 0$, then T^* is universal.

<u>Proof</u>: We first define a map $f^1: T^1 \to S^1$ extending any natural transformation $f^o: T^o \to S^o$ where S is a δ-functor.

Take any $A \in \mathcal{A}$, and let $i: A \to M$ be an effacement of T^1. Let $0 \to A \xrightarrow{i} M \to \bar{A} \to 0$ be exact. Then we have the diagram

$$\begin{array}{ccccccccc} T^o(A) & \to & T^o(M) & \overset{①}{\to} & T^o(\bar{A}) & \overset{\delta}{\to} & T^1(A) & \xrightarrow{T^1(i)} & T^1(M) \\ \downarrow f^o & & \downarrow f^o & & \downarrow f^o & & \downarrow f^1 & & \\ S^o(A) & \to & S^o(M) & \underset{②}{\to} & S^o(\bar{A}) & \overset{\delta}{\to} & S^1(A) & \longrightarrow & S^1(M) \end{array}$$

By "diagram chasing", since the top sequence is exact and $T^1(i) = 0$, there is a unique extension $f'(A): T^1(A) \to S^1(A)$. For this statement and for many later ones, we need the easy lemma:

<u>Lemma 1</u>: Consider the diagram

$$\begin{array}{ccccccc} 0 \to A' & \overset{i}{\to} & A & \overset{j}{\to} & A'' & \to 0 . \\ & & \downarrow g & & & \\ & & B & & & \end{array}$$

If the top sequence is exact and $gi = 0$, there exists a unique $h: A'' \to B$ such that $hj = g$.

Using this lemma, I shall give the above diagram chasing in detail.

The sequence $0 \to \text{Ker}\delta \to T^o(\bar{A}) \to T^1(A) \to 0$ is exact. Since T is an exact δ-functor, the row of the diagram

$$\begin{array}{c} 0 \to \text{Im } ① \overset{j}{\to} T^o(\bar{A}) \to T^1(A) \to 0 \\ \downarrow \delta f^o \\ S^1(A) \end{array}$$

is exact.

Now $\delta f^o ① = \delta ② f^o = 0$ since $\delta ② = 0$ by definition. Therefore $\delta f^o j = 0$.

Thus by the lemma, we have a unique map $f^1(A): T^1(A) \to S^1(A)$ as required.

We show f^1 is independent of the choice of M and i. Let $i: A \to M$, $i': A \to M'$ be two effacements of T^1 at A. Then

$i'' = (i,i'): A \to M'' = M + M'$ is an effacement of T^1 at A. Consider the diagram, where the rows are exact:

$$\begin{array}{ccccccccc} 0 & \to & A & \xrightarrow{i} & M & \to & \overline{A} & \to & 0 \\ & & = \downarrow & & \downarrow \overline{j} & & \downarrow r & & \\ 0 & \to & A & \xrightarrow[i'']{} & M'' & \xrightarrow[j]{} & \overline{A}'' & \to & 0 \end{array}$$

Here $\overline{j}$ is the injection, so that the first square is commutative. $j\,\overline{j}\,i = ji'' = 0$. Therefore there exists a unique $r:\overline{A} \to \overline{A}''$ making the last square commutative. This map of exact sequences gives a 3-dimensional diagram:

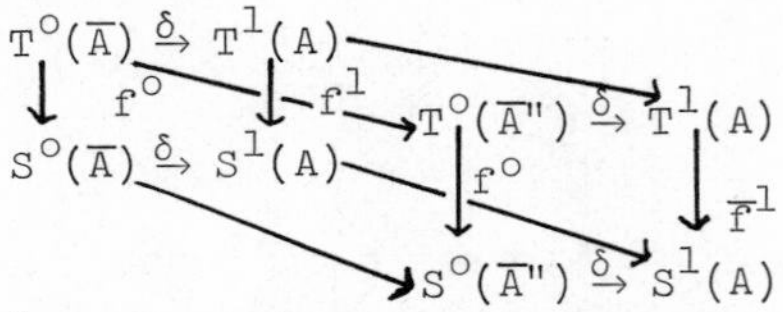

Let $\tilde{f}^1$, $\overline{f}^1$ be obtained from M' and M" respectively, by the same constructions as before. All faces of the cube are commutative except possibly that involving f^1 and $\overline{f}^1$. Therefore this face is commutative on $\mathrm{Im}(T^o(\overline{A}) \xrightarrow{\delta} T^1(A))$. But this map is onto. Therefore $f^1 = \overline{f}^1$. Similarly $\tilde{f}^1 = \overline{f}^1$, therefore $\tilde{f}^1 = f^1$.

We show f^o, f^1 preserves δ. Let $0 \to A' \xrightarrow{i} A \xrightarrow{j} A'' \to 0$ be exact, and $k:A \to M$ an effacement of T^1 at A. Then $ki:A' \to M$ is an effacement of T^1 at A' since $T(ki) = T(k)T(i)=$ and ki is mono. Consider the map of exact sequences in which we construct the map $A'' \to \overline{A}''$ by the usual method.

$$\begin{array}{ccccccccc} 0 & \to & A' & \xrightarrow{i} & A & \xrightarrow{j} & A'' & \to & 0 \\ & & = \downarrow & & k\downarrow & & \downarrow & & \\ 0 & \to & A' & \to & M & \to & \overline{A}'' & \to & 0 \end{array}$$

This map induces a cubical diagram in which we have to show the top fact commutes:

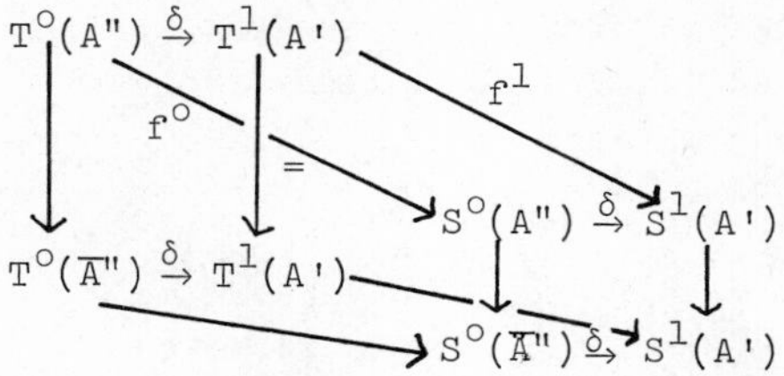

The bottom face of this commutes since it is derived from an effacement of A'. The other faces commute (except the top one) by naturality and such like. Therefore any paths linking the opposite vertices $T^o(A'')$ and $S^1(A')$ of the cube give the same map. As the vertical map $S^1(A') \to S^1(A')$ is an identity, the top face commutes.

We show naturality; i.e., let $g:A \to B$ be a map and $i:A \to M$, $j:B \to N$ be effacements of T^1; we must show that the following diagram commutes:

$$\begin{array}{ccc} T^1(A) & \xrightarrow{T^1(g)} & T^1(B) \\ \downarrow & & \downarrow f^1 \\ S^1(A) & \xrightarrow{S^1(g)} & S^1(B) \end{array}$$

Now (i, jg): $A \to M + N = P$ is an effacement of T^1 at A. We have a map of exact sequences

$$\begin{array}{ccccccccc} 0 & \to & A & \to & M + N & \to & \overline{A} & \to & 0 \\ & & g\downarrow & & \downarrow & & \downarrow & & \\ 0 & \to & B & \to & N & \to & \overline{B} & \to & 0 \end{array}$$

where $M + N \to N$ is the projection. We get another cube, and the proof goes rather similarly to the earlier ones.

We construct the transformations f^2, f^3, ... in an entirely similar manner.

<u>Corollary</u>: A cohomological δ-functor is universal as an augmented δ-functor; i.e., let $U^* = U^i{}_{o \leq i < a}$, be a cohomological δ-functor over R, $V^* = V^i{}_{o \leq i < a}$, an augmented δ-functor over S, and let $f:R \to S$ be a natural transformation. Then there is a unique map $U^* \to V^*$ of augmented δ-functors.

Here the composition $U^o \xrightarrow{\epsilon^{-1}} R \xrightarrow{f} S \xrightarrow{\epsilon} V^o$ starts the construction.

§2 Injective objects, complexes, resolutions

Definition: An object $I \in \mathcal{A}$ is called injective if for any diagram with row exact

$$\begin{array}{ccc} 0 \to & A & \xrightarrow{g} B \\ f & \downarrow & \\ & I & \end{array}$$

there is an $h: B \to I$ making the diagram commute.

Remark 2.1: I is injective $\Rightarrow$ whenever $0 \to I \to M$ is exact, I is a direct summand of M. For the completion of the diagram

$$\begin{array}{ccc} 0 \to & I & \to M \\ = & \downarrow & \\ & I & \end{array}$$

is the required projection.

Assumption 2.2: Assume that every object in $\mathcal{A}$ can be embedded in an injective one; i.e., if $A \in \mathcal{A}$, there is a mono $i: A \to I$ with I injective. This is true for modules, protosheaves, and sheaves.

Then the converse of our above remark is also true. Let $I \xrightarrow{i} Q \xrightarrow{p} I$ be a representation of I as the direct summand of an injective Q. Let $0 \to A \to B$ be exact, $f: A \to I$ a map. Then since Q is injective, $i \circ f$ can be factored through B.

$$\begin{array}{ccc} 0 \to & A & \to B \\ f & \downarrow & \downarrow h \\ & I & \xrightarrow{i} Q \end{array}$$

ph is the required map $B \to I$. Thus I is injective.

We also have the lemma:

Lemma 2: Let $\mathcal{M}$ be a collection of objects of $\mathcal{A}$ such that

(1) for every $A \in \mathcal{A}$, there is a mono $i: A \to M$, for some $M \in \mathcal{M}$

and (2) $M \in \mathcal{M}$, N direct summand of $M \Rightarrow N \in \mathcal{M}$.

Then I injective $\Rightarrow I \in \mathcal{M}$.

Proof: There is a mono $i: I \to M$ for some $M \in \mathcal{M}$. I is injective and so a direct summand of M. Therefore $I \in \mathcal{M}$.

Theorem 2: A functor $F: \mathcal{A} \to \mathcal{B}$ is effaceable $\Leftrightarrow F(I) = 0$

all injective I.

Proof: i. Let I be injective. There is a mono $i: I \to M$ with $F(i) = 0$. By our remark, $M = I + J$. The inclusion $F(I) \to F(I) + F(J)$ is zero. Therefore $F(I) = 0$.

ii. Assume $F(I) = 0$ all injective I. For any $A \in \mathcal{A}$, there is a mono $i:A \to I$, I injective. Certainly $F(i) = 0$.

Definition: By Hom(A,B) we mean the K-module of maps $A \to B$.

Proposition 1: For any B, Hom(,B) is left-exact as a functor from $\mathcal{A}$ to the category of K-modules. We leave the proof to the reader.

Proposition 2: I is injective $\Leftrightarrow$ Hom (,I) is exact.

The proof again is easy.

Proposition 3: A direct sum is injective $\Leftrightarrow$ each summand is injective. Proof easy.

Definition: A right-complex C over $A \in \mathcal{A}$ is a sequence

$$0 \to A \overset{\epsilon}{\to} C^0 \overset{\delta}{\to} C^1 \to C^2 \to \ldots \qquad (C^i = 0,\ i < 0)$$

such that $\delta^2 = 0$, $\delta\epsilon = 0$. δ is called the coboundary, ϵ the augmentation. Such a complex is called acyclic if the sequence is exact. It is called 0-injective (Object-Injective) if each C^i is injective. Define $Z^i(C) = \mathrm{Ker}\,(C^i \overset{\delta}{\to} C^{i+1})$. In a particular category $Z^i(C)$ will be called the set of i-dimensional cocycles of C.

Definition: If $f:A \to \overline{A}$ is a map, a map g over f of right-complexes C, and $\overline{C}$ over A, and $\overline{A}$ is a commutative diagram.

$$\begin{array}{ccccccccc} 0 \to & A & \overset{\epsilon}{\to} & C^0 & \overset{\delta}{\to} & C^1 & \overset{\delta}{\to} & C^2 & \to \ldots \\ & f\downarrow & & g^0\downarrow & & g^1\downarrow & & g^2\downarrow & \\ 0 \to & \overline{A} & \overset{\epsilon}{\to} & \overline{C}^0 & \overset{\delta}{\to} & \overline{C}^1 & \overset{\delta}{\to} & \overline{C}^2 & \to \ldots \end{array}$$

Definition: Two such maps, g and h, over f are homotopic if there are maps $s^i:C^i \to \overline{C}^{i-1}$ such that $g^i - h^i = \delta s^i + s^{i+1}\delta$. (which we write $g-h = \delta s + s\delta$). We have the fundamental theorem of homological algebra.

Theorem 3: Let C be an acyclic right complex over A, $\overline{C}$ an 0-injective right complex over $\overline{A}$, and $f:A \to \overline{A}$ any map.

Then (i) there is a map $g:C \to \overline{C}$ over f

and (ii) any such maps are homotopic.

For proof see CE, p. 78.

Definition: An acyclic right-complex C over A is called a resolution of A. C is called an injective resolution if it is 0-injective. By the fundamental theorem, an injective resolution is unique up to homotopy equivalence.

Theorem 4: There is an injective resolution over any $A \in \mathfrak{a}$. For proof see CE p. 80, though there it is proved for projective resolutions only. A curious consequence of this theorem is that if $f:S \to T$ is a natural transformation of left-exact K-functors such that $f(I)$ is an isomorphism for I injective, then f is a natural equivalence. We give the simple proof later when this fact is needed.

Definition: Right-Derived Functors. Let T be a covariant K-functor. Let $A \in \mathfrak{a}$ and let

$$0 \to A \xrightarrow{\epsilon} C^0 \xrightarrow{\delta} C^1 \xrightarrow{\delta} \cdots$$

be an injective resolution of A. Consider the complex T(C)

$$0 \to T(C^0) \xrightarrow{T(\delta)} T(C^1) \xrightarrow{T(\delta)} \cdots$$

Define the right-derived functors R^iT of T by

$$R^iT.(A) = H^i(T(C)).$$

These are functors. For if $f:A \to \overline{A}$ is a map and $C,\overline{C}$ injective resolutions of A and $\overline{A}$, by the fundamental theorem there is a map $g:C \to \overline{C}$ such that the following diagram commutes

$$\begin{array}{ccc} A & \to & C \\ f\downarrow & & \downarrow g \\ \overline{A} & \to & \overline{C} \end{array}$$

Any two such g induce the same map of homology, so we have a unique map $R^iT.(f)$.

It is easy to see that R^iT is unique up to natural isomorphisms. If C,C' are two injective resolutions of A, the maps $C \rightleftarrows C'$ which cover the identity map of A furnish these isomorphisms.

Note 1: $R^iT = 0$ for $i < 0$.

Note 2: R^iT are effaceable for $i > 0$. For if A is injective, $0 \overset{\approx}{\to} A \to A \to 0 \to \cdots$ is an injective resolution of A.

Note 3: The $R^{*}T$ form an exact covariant δ-functor. We prove this later.

§3

Starting with an exact category $\mathcal{A}$ satisfying assumption 2.2, let $\mathcal{C}(\mathcal{A})$, which we abbreviate to $\mathcal{C}$, be the category of cochain complexes of objects in $\mathcal{A}$ and of cochain maps. A complex here is an infinite sequence

$$\xrightarrow{\delta} A^{-n} \xrightarrow{\delta} \cdots \xrightarrow{\delta} A^{-1} \xrightarrow{\delta} A^{0} \xrightarrow{\delta} A^{1} \xrightarrow{\delta} A^{2} \xrightarrow{\delta} \cdots$$

with $\delta\delta = 0$. This is clearly a K-category, and is easily seen to be exact.

Let $\mathcal{C}_o$ be the full subcategory of $\mathcal{C}$ such that $C \in \mathcal{C}_o \Rightarrow C^i = 0$ for $i < 0$.

Note: By a full subcategory of a category $\mathcal{A}$ we mean a subcategory $\mathcal{B}$ of $\mathcal{A}$ in which, if $A,B \in \mathcal{B}$, the set of $\mathcal{A}$-maps $A \to B$ is the same as the set of $\mathcal{B}$ maps $A \to B$.

Let $\mathcal{C}^N$ be the full subcategory of $\mathcal{C}$ such that $C \in \mathcal{C}^N \Rightarrow C^i = 0$ for $i > N$. Let $\mathcal{C}_o^N = \mathcal{C}_o \cap \mathcal{C}^N$. These are all exact K-categories.

In each of these categories we can define injective objects, complexes, resolutions, and so on. Theorem 3 is also true. To prove this we need the following theorem.

Theorem 5: A complex $C \in \mathcal{C}$ is injective $\Leftrightarrow$

1. Each C^i is injective.
2. Each $Z^i(C)$ is a direct summand of C^i.
3. $H^i(C) = 0$ for all i if $C \in \mathcal{C}$ or $\mathcal{C}^N$; all $i \neq o$ $C \in \mathcal{C}_o, \mathcal{C}_o^N$.

Finally for every complex C there is a mono $j: C \to I$, with I an injective complex.

Proof: We prove the implication $\Leftarrow$ first.

Each $C^i = Z^i + D^i$. So C is the direct sum of complexes E_i of the form

$$\cdots \to 0 \to \cdots \to 0 \to D^i \to Z^{i+1} \to 0 \to \cdots$$

Thus it is sufficient to show E_i is injective if D^i, Z^{i+1} are injective. If $X \in \mathcal{C}$, a map $g: X \to E_i$ is completely determined by a map $g^{i+1}: X^{i+1} \to Z^{i+1}$. This is obvious from the diagram

$$\begin{array}{ccccccccc}
\cdots \to & X^{i-1} & \to & X^{i} & \to & X^{i+1} & \to & X^{i+2} & \to \cdots \\
 & & & & & \downarrow g^{i+1} & & & \\
\cdots \to & 0 & \to & D^{i} & \approxeq & Z^{i+1} & \to & 0 & \to \cdots
\end{array}$$

Therefore if Z^{i+1} is injective, so is E_i, and hence so is C.
<u>Note</u>: In the case $C \in \mathcal{C}_o$, the elementary complex E_{-1} will be of the form

$$\cdots \to 0 \to Z^{o} \to 0 \to \cdots$$

Next we prove for every complex there is a mono $j: C \to I$ and I is injective. Let $X \in \mathcal{C}$. For each i, choose a mono $j: X^i \to I^i$, I^i injective. Let I_i be the elementary complex

$$\cdots \to 0 \to I^i \overset{=}{\to} I^i \to 0 \to \cdots$$

There is a map $X \to I_i$ by

$$\begin{array}{ccccccccc}
\cdots \to & X^{i-2} & \to & X^{i-1} & \to & X^{i} & \to & X^{i+1} & \to \cdots \\
 & \downarrow & & \downarrow & & \downarrow j & & & \\
\cdots \to & 0 & \to & I^{i} & \to & I^{i} & \to & 0 & \to \cdots
\end{array}$$

Let I be the product of these complexes I_i. Then we have a mono $X \to I$. We leave to the reader the examination of the cases $\mathcal{C}^N$, $\mathcal{C}_o$, and $\mathcal{C}_o^N$. Finally, let $\mathcal{M}$ be the class of complexes satisfying (1), (2), and (3). Then $\mathcal{M}$ satisfies the conditions of lemma 2. Therefore I injective $\Rightarrow I \in \mathcal{M}$.

Let $\mathcal{E}$ be the K-category of short exact sequences of objects of $\mathcal{A}$. This is not an exact category. But $\mathcal{E}$ is a full subcategory of $\mathcal{C}_o^2$. So we can define exact sequences in $\mathcal{E}$, and the conclusions of the theorem hold in $\mathcal{C}_o^2$.

<u>Proposition 4</u>: Every short exact sequence can be embedded in an injective short exact sequence. (Here we mean injective as an object of $\mathcal{C}_o^2$.

Let $0 \to A' \to A \to A'' \to 0$ be exact.

Let $i': A' \to I'$, $i'': A'' \to I''$ be monos into injective objects.

Let $I = I' + I''$, and let

$0 \to I' \to I \to I'' \to 0$

be built of an injection and a projection.

Complete the following diagram.

$$\begin{array}{ccccc}
0 & \to & A' & \to & A \\
 & & i'\downarrow & & \downarrow k \\
0 & \to & I' & \overset{=}{\to} & I'
\end{array}$$

Then the following is a map of short exact sequences:

$$\begin{array}{ccccccccc} 0 & \to & A' & \to & A & \xrightarrow{p} & A'' & \to & 0 \\ & & {\scriptstyle i'}\downarrow & & \downarrow{\scriptstyle (k,0)} & & \downarrow{\scriptstyle i''} & & \\ 0 & \to & I' & \to & I & \to & I'' & \to & 0 \end{array}$$

To prove that injective resolutions exist in $\mathcal{E}$, we need the following lemma.

Lemma 3: (The 9-lemma)

In an exact category $\mathcal{A}$ let the following be a commutative diagram with exact rows.

$$\begin{array}{ccccccccc} & & (1) & & (2) & & (3) & & \\ & & 0 & & 0 & & 0 & & \\ & & \downarrow & & \downarrow & & \downarrow & & \\ 0 & \to & A' & \to & A & \to & A'' & \to & 0 \\ & & \downarrow & & \downarrow & & \downarrow & & \\ 0 & \to & B' & \to & B & \to & B'' & \to & 0 \\ & & \downarrow & & \downarrow & & \downarrow & & \\ 0 & \to & C' & \to & C & \to & C'' & \to & 0 \\ & & \downarrow & & \downarrow & & \downarrow & & \\ & & 0 & & 0 & & 0 & & \end{array}$$

Then (a) Columns (1) and (2) exact $\Rightarrow$ column (3) exact, and

(b) Columns (2) and (3) exact $\Rightarrow$ column (1) exact.

We also have

(c) Columns (1) and (3) exact and compositions in column (2) are zero $\Rightarrow$ column (2) exact.

We leave the proof to the reader.

Corollary:

If $E, E' \in \mathcal{E}$, and $\varphi: E \to E'$ is a map, then

(a) φ mono $\Rightarrow$ Coker $\varphi \in \mathcal{E}$

and (b) φ epi $\Rightarrow$ Ker $\varphi \in \mathcal{E}$.

The proof is an easy application of the 9-lemma.

The standard process of building up an injective resolution now works in $\mathcal{E}$.

Theorem 6: If T is a covariant K-functor, the right derived functors form an exact covariant δ-functor.

Proof: Let $0 \to A' \to A \to A'' \to 0$ be exact.

Let $0 \to C' \to C \to C'' \to 0$ be an injective resolution in $\mathcal{E}$ of this exact sequence.

Now each C'^i, C^i, C''^i is injective.

Therefore the sequence $0 \to C'^i \to C^i \to C''^i \to 0$ splits. Any K-functor preserves split exact sequences. So

$0 \to T(C') \to T(C) \to T(C'') \to 0$ is exact. By a standard homology argument, we obtain an exact sequence

$$\to H^i(T(C')) \to H^i(T(C)) \to H^i(T(C'') \xrightarrow{\delta} H^{i+1}(T(C')) \to \ldots,$$

i.e., an exact sequence

$$\to R^iT.(A') \to R^iT.(A) \to R^iT.(A'') \xrightarrow{\delta} R^iT.(A') \to \ldots .$$

The naturality of δ follows from the homology construction. (See CE, lemma 3.3, p.40).

Note: The construction of δ for an exact category follows from the following lemma.

Lemma 4: (The 12-Lemma)

In an exact category $\mathcal{A}$ let

$$\begin{array}{ccccccc}
 & & 0 & & 0 & & 0 \\
 & & \downarrow & & \downarrow & & \downarrow \\
 & & A & \xrightarrow{i} & B & \xrightarrow{j} & C \\
 & & \downarrow & & \downarrow & & \downarrow \\
 & & * & \to & * & \to & * & \to 0 \\
 & & \downarrow & & \downarrow & & \downarrow \\
0 & \to & * & \to & * & \to & * \\
 & & \downarrow & & \downarrow & & \downarrow \\
 & & A' & \xrightarrow{i'} & B' & \xrightarrow{j'} & C' \\
 & & \downarrow & & \downarrow & & \downarrow \\
 & & 0 & & 0 & & 0
\end{array}$$

(The $*$'s represent some objects of $\mathcal{A}$)

be a commutative diagram with the columns and the middle two rows exact. Then there is a map $k:C \to A'$ such that the following sequence

$A \xrightarrow{i} B \xrightarrow{j} C \xrightarrow{k} A' \xrightarrow{i'} B' \xrightarrow{j'} C'$ is exact, k is unique and natural w.r.t. maps of these diagrams.

We can also use the 12-Lemma to prove the following lemma.

Lemma 5: Let $0 \to C' \xrightarrow{i} C \xrightarrow{j} C'' \to 0$ be a short exact sequence of cochain complexes. Let $Z^n(C)$, as usual, denote the set of n-cocycles of C.

Then there is exact sequence

$$0 \to Z^n(C') \xrightarrow{i} Z^n(C) \to Z''(C'') \to H^{n+1}(C') \xrightarrow{i_*} H^{n+1}(C) \xrightarrow{j_*} H^{n+1}(C'') \xrightarrow{\delta} H^{n+1}(C').$$

Note that this sequence is the ordinary cohomology sequence from $H^{n+1}(C')$ on.

Remark: The set R^* could be considered as a functor from the category of functors and natural transformations to the category of δ-functors and maps of δ-functors. We have:

Proposition 5: R^*T is an augmented δ-functor over T. We define a map $\epsilon: T \to R^oT$ as follows:

Let $0 \to A \overset{\epsilon}{\to} I$ be an injective resolution of A. The map $T(\epsilon): T(A) \to T(I^o)$ defines, since $\delta\epsilon = 0$, a map $T(A) \to Z^oT(I) \approx H^oT(I) = R^oT.(A)$. Let ϵ (A) be this composition $T(A) \to R^oT.(A)$. ϵ is obviously a natural transformation.

Proposition 6: ϵ is an isomorphism $\Leftrightarrow$ T is left-exact.

i. Suppose $\epsilon(A)$ is an isomorphysm for each $A \in \mathcal{a}$. To prove the proposition we need only show R^oT is left-exact.

Let $0 \to A' \to A \to A'' \to 0$ be exact. Then the sequence $0 \to R^oT.(A') \to R^oT.(A) \to R^oT(A'') \to R^1T.(A) \to \ldots$ is exact; i.e., R^oT is left-exact.

ii. Suppose T is left-exact.

Let $0 \to A \overset{\epsilon}{\to} I$ be an injective resolution of A. Then $0 \to T(A) \xrightarrow{T(\epsilon)} T(I^o) \to T(I^1)$ is exact; i.e., $T(A) \approx Z^o(T.(I)) \approx H^o(T(I)) = R^oT(A)$. Therefore ϵ is an isomorphism.

Corollary 1: If T is left-exact, R^*T is a cohomological δ-functor over T.

Corollary 2: If T is left-exact, any cohomological δ-functor over T is naturally equivalent to R^*T.

What we need now is an easier method of computation for the right-derived functors. Injective resolutions suffer from the disadvantage that injective objects are usually too big. In particular we wish to know for what objects M and functors T does $R^iT(M) = 0$. For sheaves, with $T = \Gamma_\Phi$, this question will be gone into in great detail later on. For the present, we note the following proposition and lemma.

Proposition 7: Let T be left-exact and $0 \to A \to C$ an acyclic complex over A such that $R^iT.(C^j) = 0$ $i > o$, all j. Such a complex we call a T-resolution of A.

Then there is a natural isomorphism

$$H^iT(C) \approx R^iT.(A)$$

which preserves ϵ and δ;

i.e., if $0 \to A \to C$, $0 \to A' \to C'$ are T-resolutions and

$$\begin{array}{ccc} A & \to & C \\ \downarrow & & \downarrow \\ A' & \to & C' \end{array}$$

is a map of augmented complexes, then the following diagrams commute:

$$\begin{array}{ccccccc} & \overset{T(\epsilon)}{\nearrow} & H^{o}T(C) & & H^{i}T(C) & \overset{\approx}{\to} & R^{i}T.(A) \\ T(A) & & \downarrow \approx & & \downarrow & & \downarrow \\ & \underset{\epsilon}{\searrow} & R^{o}T.(A) & & H^{i}T(C') & \overset{\approx}{\to} & R^{i}T.(A') \end{array}$$

The reader will easily see what preservation of δ means.

Since $0 \to A \to C$ is acyclic, it is built up out of exact sequences

$$0 \to A \to C^{o} \to Z^{1} \to 0$$
$$0 \to Z^{1} \to C^{1} \to Z^{2} \to 0$$
$$\cdots\cdots$$
$$0 \to Z^{j} \to C^{j} \to Z^{j+1} \to 0$$

Applying $R^{*}T$, we have an exact sequence

$$\to R^{i-1}T.(C^{j}) \to R^{i-1}T.(Z^{j+1}) \to R^{i}T.(Z^{j}) \to R^{i}T.(C^{j}) \to \ldots$$

By assumption $R^{i}T(C^{j}) = 0$. Therefore $R^{i}T.(Z^{j}) \approx R^{i-1}T.(Z^{j+1})$. Therefore $R^{i}T(A) = R^{i}T.(Z^{o}) \approx R^{i-1}T.(Z^{1}) \approx \ldots \approx R^{1}T.(Z^{i-1})$.

From the exact sequence

$$R^{o}T.(C^{i-1}) \overset{\tau}{\to} R^{o}T.(Z^{i}) \to R^{1}T.(Z^{i-1}) \to R^{1}T.(C^{i-1}) = 0,$$

$R^{1}T.(Z^{i-1}) \approx \operatorname{Coker} \tau$.

Consider the commutative diagram, with exact row

$$\begin{array}{ccccccc} & & & & C^{i-1} & & \\ & & & \overset{\delta}{\swarrow} & \downarrow \delta & & \\ 0 & \to & Z^{i} & \to & C^{i} & \overset{\delta}{\to} & C^{i+1} \end{array}$$

This gives a diagram with exact row

$$\begin{array}{ccccccc} & & & & R^{o}T.(C^{i-1}) & & \\ & & & \overset{\tau}{\swarrow} & \downarrow \overline{\delta}^{i-1} & & \\ 0 & \to & R^{o}T.(Z^{i}) & \to & R^{o}T.(C^{i}) & \overset{\overline{\delta}^{1}}{\to} & R^{o}T.(C^{i+1}). \end{array}$$

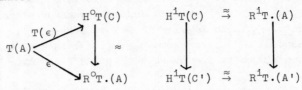

Then $\operatorname{Im} \tau \approx \operatorname{Im}\bar{\delta}^{i-1}$, and $R^oT.(Z^i) \approx \operatorname{Ker} \bar{\delta}^{\,i}$. Therefore $\operatorname{Coker} \tau \approx H^i(R^oT.(C)) \approx H^i(T(C))$, since T is left-exact. Naturality follows from the naturality of all the isomorphisms considered.

We leave the reader to prove that the isomorphism of the proposition preserves ϵ and δ.

Definition: An element $A \in \mathcal{A}$ is called T-acyclic if $R^iT.(A) = 0$ when $i > o$.

Lemma 6: Let $0 \to A \to A_1 \to \dots A_n \to B \to 0$ be exact, with $A_1,\dots,A_n$ T-acyclic. Then for $p \geq 1$, $R^pT.(B) \approx R^{p+n}T.(A)$. This is natural with respect to maps of exact sequences of the type considered.

Proof: Split the exact sequence into $0 \to A \to A_1 \to Z_2 \to 0$, $0 \to Z_2 \to A_2 \to Z_3 \to 0$, etc. and use the initial argument of the previous proposition.

Lemma 7: Let T be a covariant K-functor and let $\mathcal{M}$ be a class of objects of $\mathcal{A}$ such that:

1. I injective $\Rightarrow I \in \mathcal{M}$.
2. $0 \to M' \to M \to A \to 0$ exact and $M', M \in \mathcal{M} \Rightarrow A \in \mathcal{M}$.
3. $0 \to M' \to M \to M'' \to 0$ exact sequence of elements of $\mathcal{M} \Rightarrow 0 \to T(M') \to T(M) \to T(M'') \to 0$ exact.

Then $R^iT.(M) = 0$, $i > o$, and all $M \in \mathcal{M}$.

Proof: Let $0 \to M \overset{\epsilon}{\to} I$ be an injective resolution of M. This is built up of short exact sequences:

$$0 \to M \to I^o \to Z^1 \to 0$$
$$0 \to Z^1 \to I^1 \to Z^2 \to 0$$
$$\dots$$
$$0 \to Z^i \to I^i \to Z^{i+1} \to 0$$
$$\dots$$

Now $I^i \in \mathcal{M}$. By induction, using (2), $Z^i \in \mathcal{M}$. Therefore by (3), we get short exact sequences:

$$0 \to T(M) \to T(I^o) \to T(Z^1) \to 0$$
$$0 \to T(Z^1) \to T(I^1) \to T(Z^2) \to 0$$
$$\dots$$
$$0 \to T(Z^i) \to T(I^i) \to T(Z^{i+1}) \to 0$$
$$\dots$$

Fitting these together we get an exact sequence

$$0 \to T(M) \to T(I^o) \to T(I^1) \to T(I^2) \to \dots$$

which is just $0 \to T(M) \xrightarrow{T(\epsilon)} T(I)$.
Therefore $R^iT.(M) = 0$.

§4

This section will be of use mainly in the chapter on spectral sequences. Let T be a K-functor $\mathcal{A} \to \mathcal{B}$.

<u>Definition</u>: A natural resolution[1] N on $\mathcal{A}$ consists of:

1. a K-functor $N: \mathcal{A} \to \mathcal{C}_o(\mathcal{A})$; each N(A) is a complex

   $$\dots \to 0 \to N^o(A) \overset{\delta}{\to} N^1(A) \overset{\delta}{\to} N^2(A) \overset{\delta}{\to} \dots ,$$

   and N^i is a K-functor $\mathcal{A} \to \mathcal{A}$; and
2. a natural transformation ϵ: identity $\to N^o$; satisfying the following conditions:

   a. N is an exact functor;

   and b. the sequence $0 \to A \overset{\epsilon}{\to} N^o(A) \overset{\delta}{\to} N^1(A) \dots$ is exact.

<u>Definition</u>: A natural T-resolution N is a natural resolution such that, in addition:

c. $R^iT.N^j(A) = 0,\ i > 0,\ A \in \mathcal{A}$.

<u>Definition</u>: A resolvent functor $\mathcal{F}$ over T consists of:

1. A K-functor $\mathcal{F}: \mathcal{A} \to \mathcal{C}_o(\mathcal{B})$; each $\mathcal{F}(A)$ is a complex

   $$\dots \to 0 \to \mathcal{F}^o(A) \overset{\delta}{\to} \mathcal{F}^1(A) \overset{\delta}{\to} \mathcal{F}^2(A) \to \dots;$$

   and $\mathcal{F}^i$ is a K-functor $\mathcal{A} \to \mathcal{B}$;
2. a natural transformation $\epsilon: T \to \mathcal{F}^o$;

satisfying the following conditions:

i. $\mathcal{F}$ is an exact functor;

ii. $\epsilon: T(A) \approx H^o\,\mathcal{F}(A)$, or equivalently the sequence

$$0 \to T(A) \overset{\epsilon}{\to} \mathcal{F}^o(A) \to \mathcal{F}^1(A)$$

is exact; and

iii. I injective in $\mathcal{A} \Rightarrow 0 \to T(I) \to \mathcal{F}^o(I) \to \mathcal{F}^1(I) \to \dots$ is exact.

<u>Proposition 8</u>: Let N be a natural T-resolution, where T is

---

[1] This name is due to R. Brown.

left-exact. Then T N is a resolvent functor over T.

Proof: Since T is left-exact, condition (ii) is satisfied.

If $0 \to A' \to A \to A'' \to 0$ is exact, then so is $0 \to N^j(A') \to N^j(A) \to N^j(A'') \to 0$. Therefore, $0 \to R^oT.N^j(A') \to R^oT.N^j(A) \to R^oT.N^j(A'') \to R^1T.N^j(A') \to \ldots$ is exact. But $R^oT = T$ and $R^1T.N^j(A') = 0$. Therefore T N is an exact functor.

Let I be injective in $\mathcal{A}$. We must show that $0 \to T(I) \xrightarrow{T(\epsilon)} T\, N^o(I) \xrightarrow{T(\delta)} T\, N^1(I) \xrightarrow{T(\delta)} \ldots$ is acyclic. It is acyclic at $T\, N^o(I)$ since T is left-exact. For the other terms, we note that by proposition 7, p. 51, $H^iT\, N(I) \approx R^iT.(I) = 0$ for $i > 0$, since I is injective.

Proposition 9: If T is left-exact and $\mathcal{F}$ is a resolvent functor over T, then $H^* \mathcal{F}$ is a cohomological δ-functor over T.

Proof: If $0 \to A' \to A \to A'' \to 0$ is exact so is $0 \to \mathcal{F}(A') \to \mathcal{F}(A) \to \mathcal{F}(A'') \to 0$; taking cohomology, we have an exact δ-functor. This is augmented and ϵ is, by (ii), a natural equivalence. If I is injective, $H^i \mathcal{F}(I) = 0$, for $i > 0$, by (iii), and so $H^i \mathcal{F}$ is effaceable.

Corollary: Under the conditions of proposition 9, there is a natural isomorphism of augmented δ-functors $H^* \mathcal{F} \approx R^*T$.

Definition: Let N, N' be natural resolutions. A map of natural resolutions $g: N \to N'$ is a natural transformation of functors such that if $A \in \mathcal{A}$ the following diagram commutes:

$$\begin{array}{ccc} & & N^o(A) \\ & \overset{\epsilon}{\nearrow} & \\ A & & \downarrow\ g^o(A) \\ & \underset{\epsilon}{\searrow} & \\ & & N'^o(A) \end{array}$$

Definition: Let $\mathcal{F}$, $\mathcal{F}'$ be resolvent functors over S, T, respectively. A map of resolvent functors $g: \mathcal{F} \to \mathcal{F}'$ over a natural transformation $f: S \to T$ is a natural transformation of functors such that the following diagram commutes:

$$\begin{array}{ccc} S(A) & \xrightarrow{\epsilon} & \mathcal{F}^o(A) \\ f(A)\ \downarrow & & \downarrow\ g^o(A) \\ T(A) & \xrightarrow{\epsilon} & \mathcal{F}'^o(A) \end{array}$$

Lemma 8: If T is left-exact, N, N' are natural T-resolutions, and $g: N \to N'$ is a map of natural resolutions, then

$T\,g: T\,N \rightarrow T\,N'$ is a map of resolvent functors.

The proof is easy.

IV. THE COHOMOLOGY OF SHEAVES

§1

A cohomology theory for sheaves is a cohomological δ-functor over Γ_Φ from the category of (K-) sheaves to the category of (K-) modules. By the results of the last chapter, such a homology theory exists and is unique up to natural equivalence. We can thus talk about <u>the</u> cohomology theory of sheaves.

<u>Definition</u>: $H^i_\Phi(X,\ \ ) = R^i\Gamma_\Phi(\ \ )$.

A partial cohomology theory for sheaves is an augmented δ-functor over Γ_Φ. There is a unique map (as a map of augmented δ-functor over Γ_Φ) from the cohomology theory to any partial cohomology theory.

Note that $H^0_\Phi(X,F) \approx \Gamma_\Phi(F)$ naturally.

For each sheaf A, $Hom_\Phi(A,\ \ )$ is a covariant, left-exact K-functor from sheaves to modules.

<u>Definition</u>: $Ext^i_\Phi(A,\ \ ) = R^iHom_\Phi(A,\ \ )$.

$Ext^i_\Phi(\ \ ,\ \ )$ is in fact a functor of two variables, contravariant in the first, covariant in the second. For if $f: A \to A'$ is a map, $Hom(f,1)$. $Hom_\Phi(A',\ \ ) \to Hom_\Phi(A,\ \ )$ is a natural transformation of functors. So we have a map $Ext^i_\Phi(f,1)\colon Ext^i_\Phi(A',\ \ ) \to Ext^i_\Phi(A,\ \ )$.

<u>Remark 1</u>: $Ext^i_\Phi(A,B) = \begin{cases} 0 & i < 0 \\ Hom_\Phi(A,B) & i = 0. \end{cases}$

<u>Remark 2</u>: For fixed A, $Ext^x_\Phi(A,\ \ )$ is an exact covariant δ-functor.

<u>Remark 3</u>: For fixed B, $Ext^x_\Phi(\ \ ,B)$ is an exact contravariant δ-functor.

<u>Proof</u>: Let $0 \to A' \to A \to A'' \to 0$ be exact and let $0 \to B \to I$ be an injective resolution of B. Then we have an exact sequence of cochain complexes

$$0 \to Hom_\Phi(A'',I) \to Hom_\Phi(A,I) \to Hom_\Phi(A',I) \to 0.$$

(Here $Hom_\Phi(A,I)$, stands for the cochain complex whose objects are $Hom_\Phi(A,I^i)$, and similarly for the others.) On taking homology, we have an exact sequence:

$$\dots \to Ext^i_\Phi(A'',B) \to Ext^i_\Phi(A,B) \to Ext^i_\Phi(A',B) \xrightarrow{\delta} Ext^i_\Phi(A'',B) \to \dots$$

Remark 4: B injective $\Rightarrow Ext^i_\Phi(A,B) = 0 \quad i > 0$, all A.

Remark 5: A projective $\Rightarrow Ext^i_\Phi(A,B) = 0 \quad i > 0$, all B.

§2

We now show how a cohomology theory can be defined without using injective sheaves. The main advantage of this approach is that we can construct a natural Γ_Φ-resolution for sheaves, whereas injective resolutions cannot easily be given in this way. By proposition III. 8 and the corollary to proposition III. 9 we can use any natural Γ_Φ-resolution to calculate cohomology; its existence will be of great use in constructing the spectral sequences.

Definition: A sheaf F is flabby[1] if every section over an open set of X extends to a section over X.

Theorem 1: Let $\mathfrak{M}$ be the class of flabby sheaves. Then $\mathfrak{M}$ satisfies the conditions of lemma III 7 for $T = \Gamma_\Phi$ and any family of supports Φ.

Corollary: F flabby $\Rightarrow H^i_\Phi(X,F) = 0 \quad i > 0$ (i.e., F is Γ_Φ-acyclic, which we abbreviate to F is Φ-acyclic). In fact $\mathfrak{M}$ satisfies the following condition stronger than (3).

(3') If $0 \to M' \xrightarrow{i} A \xrightarrow{j} B \to 0$ is exact and $M' \in \mathfrak{M}$ then

[1]The French term flasque for this kind of sheaf is due to Godement. "Flabby" is a literal translation of this, and has about it the same sort of feeling.

$$0 \to \Gamma_\Phi(M') \to \Gamma_\Phi(A) \to \Gamma_\Phi(B) \to 0 \text{ is exact.}$$

We first prove (1); i.e. that all injective sheaves are in $\mathfrak{m}$.

1. Let I be injective. Then the sequence

$$0 \to \mathrm{Hom}(K_{X-U}, I) \to \mathrm{Hom}(K, I) \to \mathrm{Hom}(K_U, I) \to 0 \text{ is exact.}$$

But $\mathrm{Hom}(K_U, I) \approx \Gamma(U, I)$ naturally. Therefore the map $\rho_U^X : \Gamma(X, I) \to \Gamma(U, I)$ is epi. i.e. a section over U extends to a section over X.

(3') Let $s \in \Gamma_\Phi(B)$. Consider all sections t of A over open sets U and such that $jt = s \mid U$. We define an ordering in this set by $t_o > t_1$, if t_o is an extension of t_1.

Any chain $\{t_i\}$ has an upper bound. For if t_i is defined on U_i, define t on $\bigcup U_i$ by $t \mid U_i = t_i$.

Start with the zero section $0 \in \Gamma(X - |s|, A)$ and take a maximal $t > 0$. Certainly $|t| \subset |s| \in \Phi$ and so $|t| \in \Phi$. Therefore $t \in \Gamma_\Phi(U, A)$ for some open U.

Suppose $U \neq X$, then let $x \notin U$. Certainly, since j is onto, there is a neighborhood N_x of x and a section $r \in \Gamma(N_x, A)$ such that $jr = s \mid N_x$. Let $r \mid (N_x \cap U) - t \mid (N_x \cap U) = \sigma$ and $\in \Gamma(N_x \cap U, A)$. Then $j\sigma = 0$. By the left-exactness of $\Gamma(U, \quad)$, there is a section u of M' such that $iu = \sigma$. Extend u to X (by the flabbiness of M') and then let w' be the restriction of this section to N_x. Finally, let $w = r - iw' \in \Gamma(N_x, A)$. Then $jw = s \mid N_x$ and w and t agree on $N_x \cap U$. Thus t is not maximal, as we assumed.

2. Let $0 \to M' \xrightarrow{i} M \xrightarrow{j} F \to 0$ be exact with $M, M' \in \mathfrak{m}$. We must show F is flabby.

<u>Lemma 1</u>: If M' is flabby and $U \overset{\text{open}}{\subset} X$, then $M' \mid U$ is flabby. For any section of $M' \mid U$ is a section of M'.

Let $s \in \Gamma(U, F)$. The sequence $0 \to M' \mid U \to M \mid U \xrightarrow{j'} F \mid U \to 0$ is exact. By (3'), there is a $t' \in \Gamma(U, M \mid U)$ such that $j't' = s$. Extend t' to a section $t \in \Gamma(X, M)$ (M is flabby). Then jt is an extension of s to X. Therefore F is flabby.

§3

We need a few lemmas about protosheaves and sheaves.

Lemma 2: - and ~ are exact functors. It is obvious that - is an exact functor. Let $0 \to \overline{P}' \to \overline{P} \to \overline{P}'' \to 0$ be an exact sequence of protosheaves. $\Gamma(\ \ ,*)$ is left-exact. Therefore $0 \to \Gamma(\ \ ,\overline{P}') \to \Gamma(\ \ ,\overline{P}) \to \Gamma(\ \ ,\overline{P}'')$ is an exact sequence of stacks. But the last map is onto, since any section of $\overline{P}''$ can be lifted back, point by point, to a section of $\overline{P}$.

Applying $\sqcup$, we have the lemma

Lemma 3: If $\overline{P}$ is a protosheaf, $\tilde{\overline{P}}$ is flabby.
$\Gamma(U,\tilde{\overline{P}}) \approx \mathrm{Hom}(K_U,\tilde{\overline{P}}) \approx \mathrm{Hom}(\overline{K}_U,\overline{P})$. Any map $\overline{K}_U \to \overline{P}$ extends by zero to a map $\overline{K} \to \overline{P}$. Tracing this back through the isomorphisms, $\tilde{\overline{P}}$ is flabby.

Lemma 4: There is a natural inclusion $i:F \subset \tilde{\overline{F}}$, for all sheaves F, such that F_x is naturally a direct summand of $\tilde{\overline{F}}_x$; i.e., $\overline{F}$ is naturally a direct summand of $\overline{\tilde{\overline{F}}}$.

Proof: Under the isomorphism of lemma II. 3, let $i:F \to \tilde{\overline{F}}$ correspond to the identity map $\overline{F} \to \overline{F}$.

For convenience, let $G = \tilde{\overline{F}}$. We wish to define a projection $r:\overline{G} \to \overline{F}$ such that $r\overline{i}$ = identity. Let r correspond to the identity map $G \to \tilde{\overline{F}}$.

The composition $F \xrightarrow{i} \tilde{\overline{F}} \xrightarrow{\text{identity}} \tilde{\overline{F}}$ is simply i. By lemma 3, the composition $F \xrightarrow{\overline{i}} \overline{\tilde{\overline{F}}} \xrightarrow{r} \overline{F}$ is the identity.

Definition: An acyclic complex $\dots \xrightarrow{\delta} F^n \xrightarrow{\delta} F^{n+1} \xrightarrow{\delta} \dots$ splits if each Z^n is a direct summand of F^n. In this case $F^n/_{Z^n} \approx Z^{n+1}$ and the complex is isomorphic to one built up out of elementary complexes $\dots \to 0 \to Z^n \to Z^{n+1} \to 0 \to \dots$ in the following way:

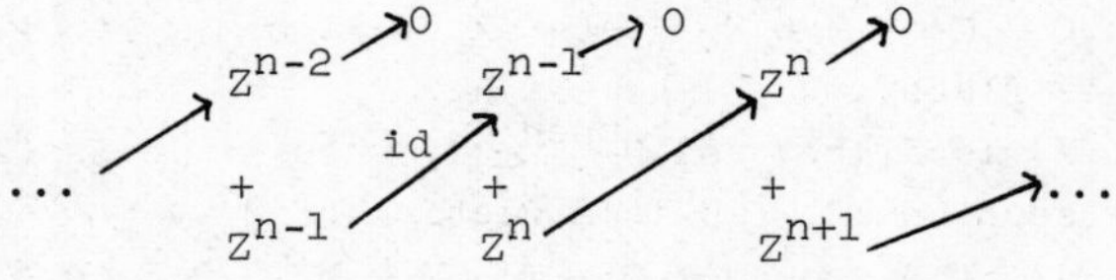

Equivalently, the complex splits if there are maps $s:F^{n+1} \to F^n$ such that $\delta s + s\delta = \mathrm{id}$. For complex $0 \to F \xrightarrow{\epsilon} C^0 \xrightarrow{\delta} C^1 \to \dots$ we have instead maps $s:C^{i+1} \to C^i$, $\eta:C^0 \to F$ such that

$$\delta s + s\delta = \mathrm{id} \quad \text{in dim} \neq 0$$

$$\text{and } s\delta = 1 - \epsilon\eta \quad \text{in dim } 0.$$

Definition: A complex $\ldots \to F^n \overset{\delta}{\to} F^{n+1} \to \ldots$ is algebraically split if the complex of protosheaves $\ldots \to \overline{F}^n \overset{\delta}{\to} \overline{F}^{n+1} \to \ldots$ splits.

Lemma 5: If $\ldots \to F^n \to F^{n+1} \to \ldots$ is algebraically split and G is any sheaf, then $\ldots \to F^n \otimes G \xrightarrow{\delta \otimes 1} F^{n+1} \otimes G \to \ldots$ is algebraically split.

Proof obvious.

Now if F is a sheaf, let $P(F) = \tilde{\overline{F}}$ and $Q(F) = \text{Coker } i = \tilde{\overline{F}}/F$. Then $0 \to F \overset{i}{\to} P(F) \to Q(F) \to 0$ is algebraically split (by lemma 4). So is $0 \to Q(F) \to PQ(F) \to Q^2(F) \to 0$. Continuing, and sticking the sequences together in the standard way, we obtain a complex

$$0 \to F \to P(F) \to PQ^1(F) \to PQ^2(F) \to \ldots$$

which is naturally, algebraically split.

Proposition 1: This complex is a natural Γ_Φ-resolution.

i. The sequence is exact, since it is algebraically split.

ii. PQ^n is a functor; moreover an exact functor. $\overline{P(F)} \approx \overline{F} + \overline{Q(F)}$ naturally. P and - are exact functors.
Therefore $\overline{Q}$ is exact. Therefore Q is exact.

iii. $PQ^n(F)$ is flabby, by lemma 3. Therefore $R^i\Gamma_\Phi . PQ^n(F) = H^i_\Phi(X, PQ^n(F)) = 0$.

This complex is called the canonical flabby resolution of F.

Remark: $\tilde{M}$, for any constant sheaf, is the sheaf of germs of functions with values in the module M.

Lemma 6: If $0 \to A \overset{i'}{\to} B \overset{j'}{\to} C \to 0$ is algebraically split there is a map

$$\begin{array}{ccccccccc} 0 \to & A & \overset{i'}{\to} & B & \overset{j'}{\to} & C & \to 0 \\ & = \downarrow & & \downarrow f & & \downarrow g & \\ 0 \to & A & \to & P(A) & \to & Q(A) & \to 0 \end{array}$$

Proof: Let $k: \overline{B} \to \overline{A}$ be a splitting of $0 \to \overline{A} \to \overline{B} \to \overline{C} \to 0$. Since $\text{Hom}(\overline{B}, \overline{A}) = \text{Hom}(B, \tilde{\overline{A}})$, k induces a map $f: B \to \tilde{\overline{A}} = P(A)$. Since $ki' = id$, $fi' \in \text{Hom}(A, \tilde{\overline{A}})$ corresponds to $id \in \text{Hom}(\overline{A}, \overline{A})$ $\ldots$ $fi' = i$.

The map g is obtained by taking quotients.

Proposition 2: Any algebraically split resolution of A can be mapped into the canonical flabby resolution of A.

Let $0 \to A \to C^o \to C^1 \to \ldots$ be the resolution.

Then $0 \to A \to C^o \to Z^1 \to 0$,

$0 \to Z^1 \to C^1 \to Z^2 \to 0$, etc. are algebraically split.

We use lemma 6 recursively to get maps

$$\begin{array}{ccccccccc} 0 & \to & A & \to & C^o & \to & Z^1 & \to & 0 \\ & & = \downarrow & & \downarrow f^o & & \downarrow g^1 & & \\ 0 & \to & A & \to & PA & \to & QA & \to & 0 \end{array}$$

and

$$\begin{array}{ccccccccc} 0 & \to & Z^i & \to & C^i & \to & Z^{i+1} & \to & 0 \\ & & \downarrow g^i & & \downarrow f^i & & \downarrow g^{i+1} & & \\ 0 & \to & Q^iA & \to & PQ^iA & \to & Q^{i+1}A & \to & 0 \end{array}$$

These fit together to give a map

$$\begin{array}{ccccccccc} 0 & \to & A & \to & C^o & \to & C^1 & \to & \ldots \\ & & = \downarrow & & \downarrow f^o & & \downarrow f & & \\ 0 & \to & A & \to & PA & \to & PQA & \to & \ldots \end{array}$$

Remark: The sheaf of Alexander-Spanier Cochains is algebraically split. For the homotopies defined in chapter II, example (vii) give the splitting.

§4 Cup products

§4.1 Let $\mathcal{A}$, $\mathcal{B}$ be exact K-categories admitting bilinear maps and tensor products. (We do not formulate exactly what this would mean; we might take $\mathcal{A}$ to be, e.g., the category of sheaves, $\mathcal{B}$ that of protosheaves, stacks, or modules.)

Let T_1, T_2, T be covariant K-functors $\mathcal{A} \to \mathcal{B}$ and assume that for $A, A' \in \mathcal{A}$ there is a natural map $T_1(A) \otimes T_2(A') \xrightarrow{\eta A,A'} T(A \otimes A')$. I.e., if $f: A \to B, g: A' \to B'$ are maps in $\mathcal{A}$, the following diagram shall commute:

$$\begin{array}{ccc} T_1(A) \otimes T_2(A') & \xrightarrow{\eta A,A'} & T(A \otimes A') \\ T_1(f) \otimes T_2(g) \downarrow & & \downarrow T(f \otimes g) \\ T_1(B) \otimes T_2(B') & \xrightarrow[\eta B,B']{} & T(B \otimes B') \end{array}$$

Let T_1^*, T_2^*, T^* be augmented δ-functors over T_1, T_2 and T respectively. A cup product for these functors (and η) shall be a natural map

$$\eta^{p,q}_{A,A'}: T_1^{\,p}(A) \otimes T_2^{\,q}(A') \to T^{p+q}(A \otimes A')$$

for each $A, A' \in \mathcal{A}$ which shall satisfy the following axioms:

I. Let p = q = 0. Then the following diagram commutes:

$$\begin{array}{ccc} T_1^0(A) \otimes T_2^0(A') & \xrightarrow{\eta^{0,0}} & T^0(A \otimes A') \\ \epsilon \otimes \epsilon \uparrow & & \uparrow \epsilon \\ T_1(A) \otimes T_2(A') & \xrightarrow{\eta} & T(A \otimes A') \end{array}$$

II. Let $0 \to A' \to A \to A'' \to 0$ be exact. Let $B \in \mathcal{A}$ and consider the commutative diagram

$$\begin{array}{ccccccc} & A' \otimes B & \to & A \otimes B & \to & A'' \otimes B & \to 0 \\ & \downarrow & & \downarrow & & \downarrow & \\ 0 \to & C' & \to & C & \to & C'' & \to 0 \end{array}$$

Then the following diagram commutes:

$$\begin{array}{ccc} T_1^p(A'') \otimes T_2^q(B) & \xrightarrow{\delta \otimes 1} & T_1^{p+1}(A') \otimes T_2^q(B) \\ \downarrow & & \downarrow \\ T^{p+q}(C'') & \xrightarrow{\delta} & T^{p+q+1}(C') \end{array}$$

The vertical maps are cup products followed by maps of the first diagram.

III. Let $0 \to B' \to B \to B'' \to 0$ be exact in $\mathcal{A}$. Let $A \in \mathcal{A}$ and consider the commutative diagram:

$$\begin{array}{ccccccc} & A \otimes B' & \to & A \otimes B & \to & A \otimes B'' & \to 0 \\ & \downarrow & & \downarrow & & \downarrow & \\ 0 \to & C' & \to & C & \to & C'' & \to 0 \end{array}$$

We use the notation w for the map $T^* \to T^*$ which is $(-1)^p$ in dimension p. Then we require the following to be commutative:

$$\begin{array}{ccc} T_1^p(A) \otimes T_2^q(B'') & \xrightarrow{w \otimes \delta} & T_1^p(A) \otimes T_2^{q+1}(B') \\ \downarrow & & \downarrow \\ T^{p+q}(C'') & \xrightarrow{\delta} & T^{p+q+1}(C') \end{array}$$

where the vertical maps are similar to those in II.

We wish to prove the theorem.

<u>Theorem 2:</u> Let T_1^*, T_2^*, T^* be cohomological δ-functors over

T_1, T_2, and T admitting cup products. Let S_1^*, S_2^* and S be augmented δ-functors over S_1, S_2 and S admitting cup products, and let $f_1: T_1 \to S_1$, $f_2: T_2 \to S_2$, and $f: T \to S$ be natural transformations such that the following diagram commutes:

$$\begin{array}{ccc} T_1(A) \otimes T_2(A') & \xrightarrow{\eta_{A,A'}} & T(A \otimes A') \\ f_1 \otimes f_2 \downarrow & & \downarrow f \\ S_1(A) \otimes S_2(A') & \xrightarrow{\xi_{A,A'}} & S(A \otimes A') \end{array}$$

Then the unique extensions $f_1^*: T_1^* \to S_1^*$, $f_2^*: T_2^* \to S_2^*$, and $f^*: T^* \to S^*$ given by theorem III preserve cup products; i.e., the following diagram commutes:

$$\begin{array}{ccc} T_1^p(A) \otimes T_2^q(A') & \to & T^{p+q}(A \otimes A') \\ \downarrow & & \downarrow \\ S_1^p(A) \otimes S_1^q(A') & \to & S^{p+q}(A \otimes A') \end{array}$$

Unfortunately, to prove this we need to make an assumption about the category $\mathcal{A}$. We assume that for any $A \in \mathcal{A}$ and each of the T_i's there is a short exact sequence $0 \to A \to P \to Q \to 0$ such that

(i) $T_i^j(P) = 0 \quad j > 0$,

(ii) for any $B \in \mathcal{A}$, $0 \to A \otimes B \to P \otimes B \to Q \otimes B \to 0$ is exact.

The exact sequence $0 \to A \to P \to Q \to 0$ may be different for T_1 and T_2. This assumption is certainly true for sheaves, when $T_1 = \Gamma_{\Phi_1}$, $T_2 = \Gamma_{\Phi_2}$, and $T = \Gamma_\Phi$ where $\Phi = \Phi_1 \cap \Phi_2$, since for any sheaf F we can take the short exact sequence to be $0 \to F \to P(F) \to Q(F) \to 0$ (see later for more details). It also works for modules over group rings, when the T^*'s are cohomologies of groups, cf. CE XII, 5.

<u>Proof of the theorem</u>: Let (p,q) denote the commutativity of the diagram of the theorem for the case p,q.

We prove (0,0), and then $(p,q) \Rightarrow (p+1,q)$ and $(p,q) \Rightarrow (p,q+1)$.

a. <u>(0,0)</u>

Consider the following cube:

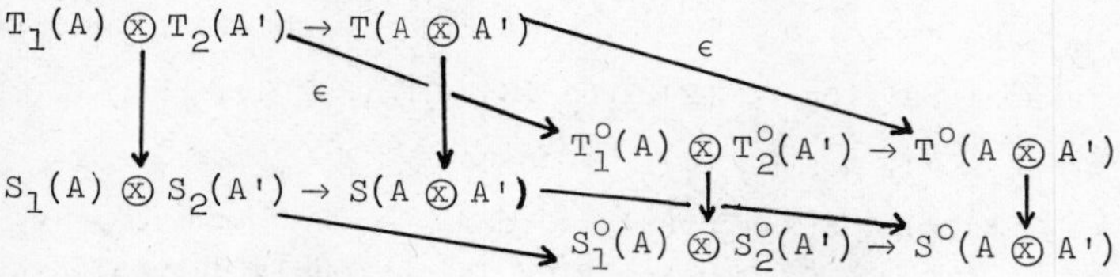

We wish to show the front face commutes. By the assumptions of the theorem, all the other faces commute. But the maps written above as ϵ are isomorphisms. Therefore the front face commutes.

b. $(p,q) \Rightarrow (p+1,q)$.

Let $0 \to A \to P \to Q \to 0$ satisfy our assumptions (i) and (ii) for T_1^*. Let $B \in \mathbf{\mathcal{A}}$. Then from the identity map of $0 \to A \otimes B \to P \otimes B \to Q \otimes B \to 0$ we obtain a cube:

$$T_1^p(Q) \otimes T_2^q(B) \xrightarrow{\delta \otimes 1} T_1^{p+1}(A) \otimes T_2^q(B)$$
$$S_1^p(Q) \otimes S_2^q(B) \to S_1^{p+1}(A) \otimes S_2^q(B)$$
$$T^{p+q}(Q \otimes B) \to T^{p+q+1}(A \otimes B)$$
$$S^{p+q}(Q \otimes B) \to S^{p+q+1}(A \otimes B)$$

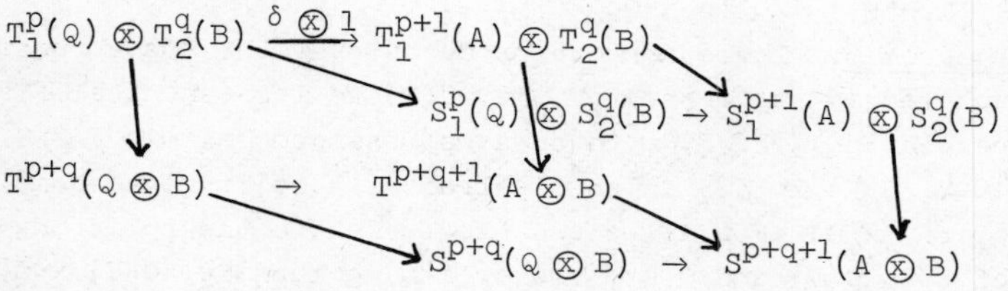

We must show the right-hand face commutes. By our assumption and by naturality and such, all the others do. Therefore the diagram commutes as far as paths starting from $T_1^p(Q) \otimes T_2^q(B)$ are concerned. But, since $T_1^1(P) = 0$ and $\otimes$ is right-exact, $\delta \otimes 1$ is epi. Therefore the right-hand face commutes.

c. $(p,q) \Rightarrow (p,q+1)$.

The argument is almost exactly the same as (b) with a map $w \otimes \delta$ replacing $\delta \otimes 1$ and with care for signs. We leave this to the reader. The theorem now follows by induction.

§4.2 Cup products for sheaves

Now let $\mathbf{\mathcal{A}}$ be the category of sheaves and $\mathbf{\mathcal{B}}$ the category of modules. Let Φ_1, Φ_2 be families of supports for X. Then $\Phi = \Phi_1 \cap \Phi_2$ is also a family of supports. Let $T_1 =$

Γ_{Φ_1}, $T_2 = \Gamma_{\Phi_2}$, and $T = \Gamma_{\Phi}$. If F,G are sheaves, there is a natural pairing $\Gamma_{\Phi_1}(F) \otimes \Gamma_{\Phi_2}(G) \to \Gamma_{\Phi}(F \otimes G)$. For if $s \in \Gamma_{\Phi_1}(F)$, $t \in \Gamma_{\Phi_2}(G)$, define $s \otimes t \in \Gamma_{\Phi}(F \otimes G)$ by $(s \otimes t)(x) = s(x) \otimes t(x)$. Clearly $|s \otimes t| \subset |s| \cap |t|$ and so $|s \otimes t| \in \Phi$. The pairing is obviously natural. We can take the T^*_α's to be $H_{\Phi_\alpha}(X, \ )$ with $\alpha = 1,2$ or nothing. Cup products would now be pairings.

$$H^p_{\Phi_1}(X,F) \otimes H^q_{\Phi_2}(X,G) \to H^{p+q}_{\Phi}(X,F \otimes G).$$

The last theorem shows that if cup products exist, they are unique; our assumption about the category $\mathcal{A}$ is satisfied by taking the short exact sequence to be $0 \to F \to P(F) \to Q(F) \to 0$. This is algebraically split, and so satisfies condition (ii). P(F) is flabby, so for any family of supports $H^p_\Phi(X,P(F)) = 0$, $p > 0$.

Theorem 3: There are cup products over the above functors and maps. For a sheaf F, let C(F) be its canonical flabby resolution. If F,G are sheaves, the complex $0 \to F \otimes G \xrightarrow{\epsilon \otimes \epsilon} C(F) \otimes C(G)$ is an algebraically split right-complex over $F \otimes G$. For if we use η's to denote the splitting maps of both C(F) and C(G), the splitting maps for $C(F) \otimes C(G)$ are $\eta \otimes \eta$ and $s \otimes 1 + \epsilon\eta \otimes s$.

Let $0 \to F \otimes G \to I$ be an injective resolution of $F \otimes G$. Then we have a map, unique up to homotopy

$$\begin{array}{ccc} & \nearrow & C(F) \otimes C(G) \\ 0 \to F \otimes G & & \downarrow \\ & \searrow & I \end{array}$$

So we have natural maps

$$\Gamma_{\Phi_1}(C(F)) \otimes \Gamma_{\Phi_2}(C(G)) \to \Gamma_{\Phi}(C(F)) \otimes C(G)) \to \Gamma_{\Phi}(I).$$

Taking cohomologies and using the natural map of the Künneth formulas which injects a product of homologies into the homology of the product, we obtain a well-defined map,

$$H^p_{\Phi_1}(X,F) \otimes H^q_{\Phi_2}(X,G) \to H^{p+q}_{\Phi}(X,F \otimes G).$$

This map is easily seen to be natural. We verify the axioms for cup products.

I. We have a cochain map

$$\begin{array}{ccc} 0 \to \Gamma_{\Phi_1}(F) \otimes \Gamma_{\Phi_2}(G) & \to & \Gamma_{\Phi_1}(C(F)) \otimes \Gamma_{\Phi_2}(C(G)) \\ \downarrow & & \downarrow \\ 0 \to \Gamma_{\Phi}(F \otimes G) & \to & \Gamma_{\Phi}(C(F) \otimes C(G)) \end{array}$$

This induces

$$\begin{array}{ccc} 0 \to \Gamma_{\Phi_1}(F) \otimes \Gamma_{\Phi_2}(G) & \to & H^{0}(\Gamma_{\Phi_1}(C(F)) \otimes \Gamma_{\Phi_2}(C(G)) \\ \downarrow & & \downarrow \\ 0 \to \Gamma_{\Phi}(F \otimes G) & \to & H^{0}_{\Phi}(C(F) \otimes C(G)) \end{array}$$

The result follows from the diagram.

$$\begin{array}{ccc} & \nearrow & \Gamma_{\Phi}(C(F) \otimes C(G)) \\ 0 \to \Gamma_{\Phi}(F \otimes G) & & \downarrow \\ & \searrow & \Gamma_{\Phi}(I) \end{array}$$

The fact that $H^{0}(\Gamma_{\Phi_1}(C(F)) \otimes \Gamma_{\Phi_2}(C(G)) \approx H^{0}_{\Phi_1}(F) \otimes H^{0}_{\Phi_2}(G)$.

II. Let $0 \to F' \to F \to F'' \to 0$ be exact. Consider the commutative diagram

$$\begin{array}{ccccccc} F' \otimes G & \to & F \otimes G & \to & F'' \otimes G & \to & 0 \\ \downarrow & & \downarrow & & \downarrow & & \\ 0 \to H' & \to & H & \to & H'' & \to & 0 \end{array}$$

If we take injective resolutions I', I, I" of H', H, H", we can, by theorem III. 3, p. 45, obtain a commutative diagram

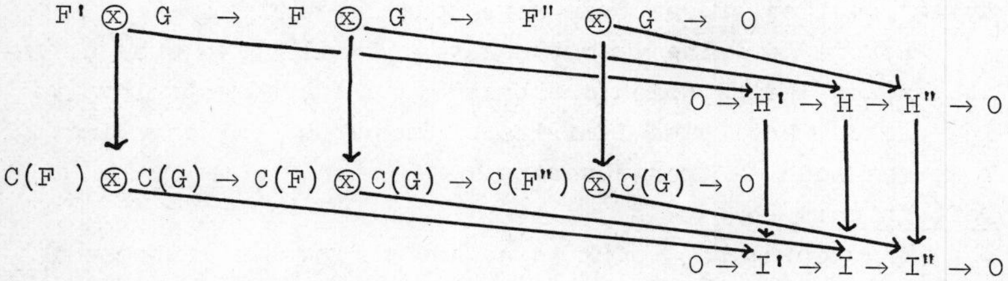

It is easily seen that this induces the correct commutativity relation for δ and $\delta \times 1$.

We leave the verification of axiom III to the reader.

V. SOME CLASSES OF Φ-ACYCLIC SHEAVES

A sheaf F is called Φ-acyclic if $H^p_\Phi(X,F) = 0$ for all $p \neq 0$. We have already met two classes of sheaves with this property, namely, injective sheaves and flabby sheaves. These are Φ-acyclic for all Φ. We now introduce a larger class of sheaves, which we will call Φ-soft. In contrast to the definition of flabby sheaves, the definition of Φ-soft sheaves will involve a specific family Φ. These sheaves will not be Φ-acyclic for all families Φ. We must assume that Φ satisfies the conditions of the following definition:

Definition:

Φ is paracompactifying (French: paracompactifiante) if

(a) Each $A \in \Phi$ has a neighborhood in Φ; i.e., there is a $B \in \Phi$ such that $A \subset \text{Int } B$, and

(b) Each $A \in \Phi$ is paracompact.

I will generally abbreviate this by writing "Φ is PF". Note that in Cartan's theory of sheaves, only paracompactifying families of supports were considered.

Before defining Φ-soft sheaves, I will give some definitions and lemmas which are basic in all arguments involving paracompactifying families of supports. By covering, I always mean an open covering.

Definition:

A covering $\{U_\alpha\}$ of X is called shrinkable if there is another covering $\{V_\alpha\}$ such that $\overline{V}_\alpha \subset U_\alpha$ for all α.

A classical theorem states that every point-finite covering of a normal space is shrinkable.

Definition:

A Φ-covering of X is a locally finite covering $\{U_\alpha \mid \alpha \in I\}$ of X such that there is a set $C \in \Phi$ and an element $\alpha_o \in I$ with the property that $U_\alpha \subset C$ for all $\alpha \neq \alpha_o$.

We refer to U_{α_o} as the exceptional set.

Remark:

U_{α_o} is unique unless $X \in \Phi$ in which case any α_o will do. In this case we assume some definite choice for α_o.

Lemma 1:

If Φ is PF, any Φ-covering is shrinkable.

Proof:

Let the notation be as in the definition of Φ-coverings. $\{U_\alpha \cap C\}$ is a locally finite covering of C. But, C is paracompact and hence normal. Therefore, we can shrink this covering to $\{V'_\alpha\}$. Define $V_\alpha = V'_\alpha$ for $\alpha \neq \alpha_o$ and $V_{\alpha_o} = V'_{\alpha_o} \cup X - C$. Then $\{V_\alpha\}$ is the required shrinking of $\{U_\alpha\}$.

Remark:

A covering obtained by shrinking a Φ-covering is obviously again a Φ-covering.

Lemma 2: Let Φ be PF.

Let $\mathfrak{M}$ be any covering of X which includes a set X - A with $A \in \Phi$. Then $\mathfrak{M}$ is refined by a Φ-covering $\mathfrak{N}$, the exceptional set being contained in X - A.

Proof:

A has a neighborhood C in Φ. For each $x \in X$, choose a neighborhood N_x such that N_x is contained in some set of $\mathfrak{M}$. If $x \in \text{Int } C$, take N_x so small that $N_x \subset \text{Int } C$. Now, $\{N_x \cap C\}$ covers C which is paracompact. Take a locally finite refinement $\mathfrak{W}$ of this covering. Throw away all sets of $\mathfrak{W}$ not contained in Int C. Then add X - A to get the required Φ-covering, the exceptional set being X - A.

This lemma is usually applied to the following situation:

We have a sheaf F over X, a covering $\{N_\beta\}$ of X and a section $s_\beta \in \Gamma(N_\beta, F)$ for each β. We assume that some N_β has the form X - A with $A \in \Phi$ and that $s_{\beta_o} = 0$.

We want to fit the sections s_β together to get a section s $\in \Gamma(X,F)$ with support in Φ. To do this, we take a Φ-covering $\{U_\alpha\}$ which refines $\{N_\beta\}$. For each α, we choose a β such that $U_\alpha \subset N_\beta$ and define $t_\alpha = s_\beta \mid U_\alpha$. Of course, we must be careful to choose N_{β_o} to be the set containing the exceptional U_{α_o}. This insures that t_{α_o} is zero. We now make use of the local finiteness of $\{U_\alpha\}$ to fit the t_α's together. The fact that $t_{\alpha_o} = 0$ while the U_α with $\alpha \neq \alpha_o$ are in C insures that the resulting section has its support contained in C. Therefore this support will be in Φ.

A simple example of this argument is given by the proof of the next lemma. In this lemma, no family of supports is used, but except for this, the argument is exactly as above.

Lemma 3:

Let X be paracompact and A a closed subset of X. Let F be any sheaf over X. Then any section of F over A can be extended to a neighborhood of A.

Proof:

Assume $s \in \Gamma(A,F)$ is given. Every point $x \in A$ has a neighborhood N_x so small that there is a section $s_x \in \Gamma(N_x,F)$ such that s_x and s agree over $N_x \cap A$. To find s_x, it is sufficient to find a section t_x over some neighborhood M_x of x such that $t_x(x) = s(x)$ and then take N_x to be $(M_x - A) \cup \{y \in A \mid t_x(y) = s(y)\}$.

For $x \notin A$, choose $N_x = X - A$ and $s_x = 0$. Then $\{N_x\}$ covers X. Take a locally finite refinement $\{U_\alpha\}$ and apply the above argument to get sections $t_\alpha \in \Gamma(U_\alpha,F)$ such that $t_\alpha \mid U_\alpha \cap A = s \mid U_\alpha \cap A$.

Shrink $\{U_\alpha\}$ to a covering $\{V_\alpha\}$ such that $\overline{V}_\alpha \subset U_\alpha$. Let W be the set of x such that the $t_\alpha(x)$ have the same value for all α such that $x \in \overline{V}_\alpha$. If $x \in W$, let t(x) be this common value, i.e., $t(x) = t_\alpha(x)$ for any α such that $x \in \overline{V}_\alpha$.

Obviously, $W \supset A$ and $t \mid A = s$. We must show W is open and t is continuous. If $x \in W$, some neighborhood meets only a finite number of U_α. A smaller neighborhood meets only those $\overline{V}_\alpha$ such that $x \in \overline{V}_\alpha$. At points y in this neighborhood, only those α such that $x \in \overline{V}_\alpha$ are used to decide whether $y \in W$ and to determine t(y). Now, these t_α agree at x and therefore in a neighborhood of x. This neighborhood is contained in W. Therefore W is open. Finally, in this neighborhood, t is obtained by piecing together a finite number of t_α's each defined over a set relatively closed in this neighborhood (i.e., the intersection of the neighborhood with a $\overline{V}_\alpha$). Therefore t is continuous.

Corollary:

Let A be closed in X and have a paracompact neighborhood in X. Let F be any sheaf over X. Then any section of F over A extends to a neighborhood of A.

Proof:

If B is a paracompact neighborhood of A, we apply the lemma with B in place of X.

We now define Φ-soft sheaves. There are four possible definitions. We will show that these all coincide if Φ is PF or, more generally, if every set of Φ has a neighborhood in Φ (the sets of Φ not being assumed paracompact).

Definition:

Let Φ be any family of supports in X. Let F be a sheaf over X. Then

(a) F is Φ-soft_1 if, whenever A is a closed set of X and s is a section of F over A with support in Φ, we can extend s to a section over X with support in Φ;

(b) F is Φ-soft_2 if, whenever $A \in \Phi$ and s is a section of F over A, we can extend s to a section over X with support in Φ;

(c) F is Φ-soft_3 if, whenever $A \in \Phi$, any section of F over A extends to a section of F over X;

(d) F is Φ-soft_4 if, whenever $A, B \in \Phi$ and $A \subset B$, any section of F over A extends to a section of F over B.

Remark:

Clearly Φ-$\text{soft}_1 \Rightarrow \Phi$-$\text{soft}_2 \Rightarrow \Phi$-$\text{soft}_3 \Rightarrow \Phi$-$\text{soft}_4$. Grothendieck defines Φ-soft to be Φ-soft_2.

Proposition 1:

Suppose every set in Φ has a neighborhood in Φ. Then all four definitions of Φ-soft agree.

Proof:

It is sufficient to show that Φ-$\text{soft}_4 \Rightarrow \Phi$-$\text{soft}_1$. Let A be closed in X and $s \in \Gamma(A,F)$ with support $|s| \in \Phi$. Let

$B \in \Phi$ be a neighborhood of $|s|$ and let $C \in \Phi$ be a neighborhood of B. The set $D = (A \cap C) \cup (C\text{-Int } B)$ is a closed subset of C and so is in Φ. We define a section s' over D by $s' \mid (A \cap C) = s \mid (A \cap C)$ and $s' \mid (C\text{-Int } B) = 0$. Then s' is obviously continuous since s is zero outside Int.B.

Extend s' to a section s" over C using the property Φ-soft$_4$. Then extend s" to a section t over X by $t \mid C = s"$, $t \mid (X - \text{Int } B) = 0$. Then t is continuous because s" is zero outside Int B. Clearly $|t| \subset B$ and so $|t| \in \Phi$. Finally, $t \mid (A \cap C) = s \mid (A \cap C)$ and $t \mid (A\text{-Int } B) = 0 = s \mid (A\text{-Int } B)$. Therefore $t \mid A = s$.

<u>Proposition 2</u>:

If Φ is PF, flabby implies Φ-soft.

<u>Proof</u>:

Let F be flabby. We show F is Φ-soft$_3$. Let s be a section of F over $A \in \Phi$. Since Φ is PF, A has a paracompact neighborhood in X. Therefore, s extends to a section s' over an open neighborhood U of A. We then extend s' to X by using the flabbiness of F.

<u>Corollary</u>:

If Φ is PF, injective implies Φ-soft.

<u>Proof</u>:

Injective always implies flabby.

<u>Proposition 3</u>:

Assume Φ is PF. Then

(1) $0 \to F' \xrightarrow{i} F \xrightarrow{j} F" \to 0$ exact and F' Φ-soft imply $0 \to \Gamma_\Phi(F') \to \Gamma_\Phi(F) \to \Gamma_\Phi(F") \to 0$ is exact and

(2) Assume $0 \to F' \to F \to F" \to 0$ is exact. Let F' and F be Φ-soft. Then F" is Φ-soft.

<u>Proof</u>:

We first show that (1) implies (2).

Let $A \in \Phi$ and $s \in \Gamma(A,F")$.

The sequence $0 \to F' \mid A \to F \mid A \to F" \mid A \to 0$ is exact and $F' \mid A$ is Φ_A-soft where $\Phi_A = \{C \in \Phi \mid C \subset A\}$. To see this, use property Φ-soft$_3$. Since Φ_A is clearly PF, we can find, using (1) $t \in \Gamma(A,F \mid A)$ such that $j(t) = s$. We now extend t to X, using the fact that F is Φ-soft. Finally, we get the required extension of s by applying j to this extension of t.

We must now prove (1). Since Γ_Φ is left-exact, we only have to show $\Gamma_\Phi(F) \to \Gamma_\Phi(F")$ is an epimorphism.

Let $s \in \Gamma_\Phi(F")$. Let $A = |s|$. For each $x \in X$, there is a neighborhood N_x and a section $s_x \in \Gamma(N_x, F)$ such that $js_x = s \mid N_x$. If $x \notin A$, we choose $N_x = X - A$ and $s_x = 0$. Applying the standard argument, we find a Φ-covering $\{U_\alpha\}$ and sections $t_\alpha \in \Gamma(U_\alpha, F)$ such that $j\, t_\alpha = s \mid U_\alpha$ and such that $t_{\alpha_o} = 0$ for the exceptional α_o.

Shrink $\{U_\alpha\}$ to a covering $\{V_\alpha\}$. Consider sets which are unions of some of the $\overline{V}_\alpha$'s. These are closed since $\{\overline{V}_\alpha\}$ is locally finite. Consider sections t over such sets such that $t \mid \overline{V}_\alpha - t_\alpha$, when defined, is a section of $F' \subset F$.

Order these t's by extension. Zorn's lemma applies because the local finiteness of $\{\overline{V}_\alpha\}$ insures that the upper bound of an increasing chain of t's will be continuous.

Start with the zero section over $\overline{V}_{\alpha_o}$ and take a maximal extension t. Suppose we can show that t is defined over all of X. Then $|t| \in \Phi$ because t is zero over the exceptional set V_{α_o} and so $|t| \subset C$ where $C \in \Phi$ is the set considered in the definition of a Φ-covering. Also, $j(t).\,(x) = j(t_\alpha +$ section of $F').\,(x) = j(t_\alpha).\,(x) = s(x)$ (where $x \in V_\alpha$). Therefore, we have only to show that t is defined over all of X.

Let D be the domain of t. Suppose $\overline{V}_{\alpha'} \not\subset D$. Then $\alpha' \neq \alpha_o$, therefore $V_{\alpha'} \subset C$. Thus $\overline{V}_{\alpha'} \in \Phi$. And so $D \cap \overline{V}_{\alpha'} \in \Phi$ also.

Now, $t \mid D \cap \overline{V}_{\alpha'} - t_{\alpha'} \mid D \cap \overline{V}_{\alpha'}$ is a section of F'. Extend it to $\overline{V}_{\alpha'}$ using the Φ-softness of F'. This gives a section r of F' over $\overline{V}_{\alpha'}$. But, $t_{\alpha'} + r$ is a section of F over $\overline{V}_{\alpha'}$ which agrees with t over $D \cap \overline{V}_{\alpha'}$. Therefore we can extend t to $D \cup \overline{V}_{\alpha'}$ by taking t over D and $t_{\alpha'} + r$ over $\overline{V}_{\alpha'}$. This gives a contradiction since t was assumed maximal.

<u>Corollary</u>:

If Φ is PF and F is Φ-soft, then F is Φ-acyclic.

<u>Proof</u>:

This follows from the proposition and the fact that injective implies Φ-soft. We use the same lemma which was used to show that flabby implies Φ-acyclic (Lemma III. 7, p.53).

If Φ is PF, another class of Φ-acyclic sheaves is the class of fine sheaves.

Definition:

A sheaf F over X is fine if, for every locally finite covering $\{U_\alpha\}$ of X, there exist endomorphisms $l_\alpha : F \to F$ such that

(a) $|l_\alpha| \subset \overline{U}_\alpha$

(b) $\Sigma l_\alpha = \text{id}$.

Note that (a) and the local finiteness of $\{U_\alpha\}$ imply that the sum Σl_α is well defined.

The existence of many fine sheaves is given by the following proposition.

Proposition 4:

If $\overline{M}$ is any protosheaf, then $\tilde{\overline{M}}$ is fine.

Proof:

Let $\{U_\alpha\}$ be a locally finite covering of X. For each x, choose some α_x such that $x \in U_{\alpha_x}$. Define endomorphisms $l'_\alpha : \overline{M} \to \overline{M}$ as follows:

If $\alpha = \alpha_x$, then $l'_\alpha | \overline{M}_x = \text{id}$.

If $\alpha \neq \alpha_x$, then $l'_\alpha | \overline{M}_x = 0$.

These l'_α's induce maps $l_\alpha : \tilde{\overline{M}} \to \tilde{\overline{M}}$ since ~ is a functor. Suppose $x \notin \overline{U}_\alpha$. Since l'_α is zero on all $\overline{M}_y$ with $y \in X - \overline{U}_\alpha$, l'_α is zero over a neighborhood of x. Therefore the same is true of l_α and so, $x \notin |l_\alpha|$. This shows $|l_\alpha| \subset \overline{U}_\alpha$. In a small neighborhood of any x, only a finite number of l'_α will be non-zero and the sum of the l'_α will be the identity. Consequently, $\Sigma l_\alpha = \text{id}$ also.

Corollary:

Every sheaf F can be imbedded in a fine sheaf. (We merely take the canonical imbedding $F \to \tilde{\overline{F}}$).

Lemma 4:

If $F \subset G$ is a direct summand of G and G is a fine then F is fine.

Proof:

Let $r : G \to F$ be a "retraction", i.e., a sheaf map such

that $r \mid F = \mathrm{id}$. Let $i: F \to G$ be the inclusions. Let $\{U_\alpha\}$ be a locally finite covering of X. Find endomorphisms $l_\alpha: G \to G$ such that $|l_\alpha| \subset \overline{U}_\alpha$ and $\Sigma l_\alpha = \mathrm{id}$. Then the maps $r\, l_\alpha\, i$: $F \to F$ obviously satisfy $|r\, l_\alpha\, i| \subset \overline{U}_\alpha$ and $\Sigma r\, l_\alpha\, i = ri = \mathrm{id}$.

Corollary:

Every injective sheaf is fine.

Proof:

Let I be injective. Imbed I in a fine sheaf G. Since I is injective, it is a direct summand of G.

Proposition 5:

If Φ is PF and F is fine, then F is Φ-soft.

Proof:

Let $s \in \Gamma(A,F)$ with $A \in \Phi$.
For each $x \in A$, there is a neighborhood N_x and a section $s_x \in \Gamma(N_x,F)$ such that $s_x \mid N_x \cap A = s \mid N_x \cap A$. For $x \notin A$, choose $N_x = X - A$ and $s_x = 0$. In the usual way, we find a Φ-covering $\{U_\alpha\}$ of X and sections $t_\alpha \in \Gamma(U_\alpha,F)$ such that $t_\alpha \mid U_\alpha \cap A = s \mid U_\alpha \cap A$ and such that $t_{\alpha_0} = 0$ for the exceptional α_0.

Shrink $\{U_\alpha\}$ to $\{V_\alpha\}$. Using the fineness of F, we get endomorphisms $l_\alpha: F \to F$ such that $|l_\alpha| \subset \overline{V}_\alpha$ and $\Sigma l_\alpha = \mathrm{id}$. Define $t(x) = \Sigma l_\alpha t_\alpha(x)$. This makes sense because t_α is defined over U_α while l_α is zero outside $\overline{V}_\alpha \subset U_\alpha$. Obviously t is continuous and $t \mid A = s$. This shows that F is Φ-soft$_3$. Note that $|t| \subset C$ (the set occurring in the definition of Φ-coverings). Therefore $|t| \in \Phi$, and so we have shown directly that F is Φ-soft$_2$.

Corollary:

If Φ is PF and F is fine, then F is Φ-acyclic.

It is possible to prove a generalization of this corollary which will be useful in the applications involving singular chains.

Definition:

Let F be a sheaf over X. Let $\mathcal{C}$ be a collection of endomorphisms of F. We say that F is $\mathcal{C}$-fine if, given any locally finite covering $\{U_\alpha\}$ of X, there are endomorphisms $l_\alpha: F \to F$ such that

(a) $|1_\alpha| \subset \overline{U}_\alpha$,
and (b) $\Sigma 1_\alpha \in \mathcal{C}$.

For example, if $\mathcal{C}$ contains only the identity map, then $\mathcal{C}$-fine is the same as fine.

Proposition 6:

Let Φ be PF.

Let F be $\mathcal{C}$-fine. Then, for $p > o$, every element of $H^p_\Phi(X,F)$ is annihilated by a map in $\mathcal{C}$. That is, if $u \in H^p_\Phi(X,F)$, there is a map $f \in \mathcal{C}$ such that $f_*(u) = o$.

Proof:

The proof is based on two lemmas.

We recall the definitions of the functors P and Q. If F is any sheaf, $P(F) = \tilde{\overline{F}}$ and $Q(F) = P(F)/F$. Note that these functors preserve supports; i.e., if $f:F \to G$, then $|P(f)| \subset |f|$ and $|Q(f)| \subset |f|$. This follows from the trivial fact that the functors "-" and "~" have this property. Obviously, any composition of P's and Q's will again have this property.

We also recall the canonical flabby resolution of F, $0 \to F \to P(F) \to PQD \to PQ^2(F) \to \ldots$ (or $PQ^*(F)$ for short) and the fact that if we apply Γ_Φ to this resolution (omitting the term F), we get a cochain complex whose cohomology groups are naturally isomorphic to $H^*_\Phi(X,F)$.

Lemma 5:

Let F be any sheaf and Φ any family of supports. Then for $p > o$, there is a natural exact sequence $\Gamma_\Phi(PQ^{p-1}(F)) \underset{j}{\to} \Gamma_\Phi(Q^p(F)) \to H^p_\Phi(X,F) \to 0$ where j is induced by the natural epimorphism $PQ^{p-1}(F) \to Q^p(F)$ given by the fact that $Q^p(F) = Q\,Q^{p-1}(F)$ is defined as a quotient of $PQ^{p-1}(F)$.

Proof:

We can factor $PQ^{p-1}(F) \to PQ^{p+1}(F)$ as follows:

$$\begin{array}{ccccc} PQ^{p-1}(F) & \to & PQ^p(F) & \to & PQ^{p+1}(F) \\ \text{epi}\searrow & & \nearrow\text{mono}\quad \text{epi}\searrow & & \nearrow\text{mono} \\ & Q^p(F) & & Q^{p+1}(F) & \end{array}$$

Since Γ_Φ is left exact, $0 \to \Gamma_\Phi(Q^p(F)) \to \Gamma_\Phi(PQ^p(F)) \to \Gamma(PQ^{p+1}(F))$ is exact. Therefore $\Gamma_\Phi(Q^p(F)) \xrightarrow{\approx} Z^p(\Gamma_\Phi(PQ^*(F))$.

Now,

$$\begin{array}{ccc} & \Gamma_\Phi(PQ^{p-1}(F)) & \\ \swarrow & & \searrow \\ \Gamma_\Phi(Q^p(F)) & \overset{\approx}{\rightarrow} & Z^p(\Gamma_\Phi(PQ^{*}F(F)) \end{array}$$

and the image of the second vertical map is $B^p\Gamma_\Phi(PQ^{*}F)$. Therefore $H^p_\Phi(X,F) = Z^p/_{B^p}$ is naturally isomorphic to the cokernel of $\Gamma_\Phi(PQ^{p-1}(F)) \to \Gamma_\Phi(Q^p(F))$.

Lemma 6:

Let S and T be covariant linear functors from sheaves on X to sheaves on X which preserve supports (in the sense mentioned above in connection with P and Q). Let $j:S \to T$ be a natural transformation.

Assume Φ is PF, F is $\mathcal{C}$-fine, and $j_F:S(F) \to T(F)$ is an epimorphism.

Then, every element of the cokernel of $\Gamma_\Phi(S(F)) \to \Gamma_\Phi(T(F))$ is annihilated by some map in $\mathcal{C}$.

Remark 1:

Proposition 6 follows immediately from this lemma with $S = P\,Q^{p-1}$ and $T = Q^p$.

Remark 2:

This lemma is a generalization of the fundamental theorem of Cartan's theory of sheaves.

Proof of the lemma:

Let $s \in \Gamma_\Phi(T(F))$ represent an element of the cokernel. Let $A = |s|$. For each $x \in X$, choose a neighborhood N_x and section $s_x \in \Gamma(N_x,S(F))$ such that $j(s_x) = s \mid N_x$. If $x \notin A$, choose $N_x = X - A$ and $s_x = 0$. In the usual way, we find a Φ-covering $\{U_\alpha\}$ and sections $t_\alpha \in \Gamma(U_\alpha,S(F))$ such that $j(t_\alpha) = s \mid U_\alpha$ and $t_{\alpha_o} = 0$ for the exceptional α_o.

Shrink $\{U_\alpha\}$ to $\{V_\alpha\}$ and find endomorphisms l_α of F such that $|l_\alpha| \subset \overline{V}_\alpha$ and $\Sigma l_\alpha = f \in \mathcal{C}$. Define $t(x) = \Sigma S(l_\alpha)t_\alpha(x)$. This makes sense because $|S(l_\alpha)| \subset |l_\alpha| \subset \overline{V}_\alpha$. t is obviously continuous and has support in Φ because $t_{\alpha_o} = 0$ and so $|t| \subset C$ (the set used in defining Φ-coverings).

But, $j(t)(x) = \Sigma j\, S(l_\alpha)t_\alpha(x) = \Sigma T(l_\alpha)jt_\alpha(x) = \Sigma T(l_\alpha)\, s(x) = T(f)\, s(x)$.

Therefore $\Gamma_\Phi T(f).(s)$ is in the image of $\Gamma_\Phi(j)$. In other words, f annihilates the element of the cokernel represented by s.

VI. THE SECTIONS OF SHEAVES

The general problem we will consider here is the following:

Let $\underline{S}$ be a stack with "restriction" maps φ_V^U and $S = L(\underline{S})$ its sheaf. We then have the natural map $\varphi:\underline{S}(X) \to \Gamma(X,S)$. The problem is first to define supports in $\underline{S}(X)$ so that φ preserves supports, then to find the kernel of φ, and finally to decide when $\varphi:\underline{S}_\Phi(X) \to \Gamma_\Phi(S)$ is an epimorphism. (See below for definition of $\underline{S}_\Phi$.)

The first two parts of the problem are quite easily solved.

Definition:

Let $\underline{S}$ be a stack over X. If $s \in \underline{S}(U)$, define the support $|s|$ of s as follows:

A point $x \in U$ does not belong to $|s|$ if and only if x has an open neighborhood N_x such that $\varphi_{N_x}^U(s) = 0$.

Obviously $|s|$ is closed in U because $x \notin |s|$ implies that all points in N_x are not in $|s|$.

Lemma 1:

Let S be the sheaf of $\underline{S}$ and $\varphi:\underline{S}(U) \to \Gamma(U,S)$ be the natural map. Then $|\varphi(s)| = |s|$ for all $s \in \underline{S}(U)$.

Proof:

1. If $x \notin |s|$, then $\underline{S}(U) \to \underline{S}(N_x) \to S_x$ annihilates s. Consequently, $\varphi(s).(x) = 0$ and so $x \notin |\varphi(s)|$.
2. If $x \notin |\varphi(s)|$, then $\underline{S}(U) \to S_x$ annihilates s. Since S_x is the direct limit of $\underline{S}(N_x)$ over neighborhoods of x, there is an N_x such that $\underline{S}(X) \to \underline{S}(N_x)$ annihilates s. Therefore $x \notin |s|$.

Corollary:

The kernel of $\varphi:\underline{S}(U) \to \Gamma(U,S)$ is $\{s \mid |s| = \phi\}$. Such s are called locally zero elements.

This corollary follows immediately from the lemma and the observation that an element $t \in \Gamma(U,S)$ is zero if and only if $|t| = \phi$.

Definition:

If Φ is a family of supports in X, let $\underline{S}_{\Phi}(X) = \{s \in \underline{S}(X) \mid |s| \in \Phi\}$.

It follows from lemma 1 that $\varphi: \underline{S}_{\Phi}(X) \to \Gamma_{\Phi}(S)$ and the kernel of φ consists of locally zero elements.

The final part of our problem is to decide when this map is an epimorphism. We note first of all that it is not necessary to worry about supports in solving this problem. That is, suppose $s \in \Gamma_{\Phi}(S)$, and suppose we can find $t \in \underline{S}(X)$ such that $\varphi(t) = s$. It then follows automatically that $|t| = |\varphi(t)| = |s| \in \Phi$. Therefore $t \in \underline{S}_{\Phi}(X)$.

Suppose now that we are given $s \in \Gamma_{\Phi}(S)$. For each $x \in X$, there is an open neighborhood N_x and an element $s_x \in \underline{S}(N_x)$ such that $\varphi(s_x) = s \mid N_x$. If $x \notin |s|$, we can obviously choose $N_x = X - |s|$ and $s_x = 0$. We now consider either the covering $\{N_x\}$ itself or a suitable refinement. In this way we get a covering $\{U_\alpha\}$ of X and sections $s_\alpha \in \underline{S}(U_\alpha)$ such that $\varphi(s_\alpha) = s \mid U_\alpha$. This last condition implies that the differences

$$\varphi^{U_\alpha}_{U_\alpha \cap U_\beta}(s_\alpha) - \varphi^{U_\beta}_{U_\alpha \cap U_\beta}(s_\beta)$$

are locally zero. We want to piece together the s_α to get an element $t \in \underline{S}(X)$ such that $\varphi^{X}_{U_\alpha}(t) - s_\alpha$ is locally zero. This is usually much easier to do if we assume that the differences

$$\varphi^{U_\alpha}_{U_\alpha \cap U_\beta}(s_\alpha) - \varphi^{U_\beta}_{U_\alpha \cap U_\beta}(s_\beta)$$

are zero and not just locally zero.

Definition:

We say a stack $\underline{S}$ has the collation property if the following property holds for all coverings $\{U_\alpha\}$ of X: Suppose we have elements $s_\alpha \in \underline{S}(U_\alpha)$ such that

$$\varphi^{U_\alpha}_{U_\alpha \cap U_\beta}(s_\alpha) = \varphi^{U_\beta}_{U_\alpha \cap U_\beta}(s_\beta)$$

for all α, β. Then there is an element $t \in \underline{S}(X)$ such that

$\varphi^X_{U_\alpha}(t) = s_\alpha$ for all α.

If we weaken the conclusion to read "$\varphi^X_{U_\alpha}(t) - s_\alpha$ is locally zero for all α," we then say that S has the approximate collation property.

Finally, if we assume the above property holds only for coverings of a certain class (e.g., locally finite coverings or Φ-coverings), we say S has the collation property (or approximate collation property) with respect to this class.

The next proposition follows immediately from the above definition and the remarks preceding it.

Proposition 1:

Let S be a stack with the collation property. Assume that for every U, S(U) has no locally zero elements except zero. Then $\underline{S}_\Phi(X) \to \Gamma_\Phi(S)$ is an epimorphism (and, in fact, an isomorphism).

Remark:

If S(X) is a stack over X and U is open in X, we can define a stack $\underline{S} \mid U$ over U by considering only those S(V) for $V \subset U$. If S is the sheaf of S, then the sheaf of $\underline{S} \mid U$ is clearly $S \mid U$. The above proposition shows that a necessary and sufficient condition for S to be the stack of its sheaf (i.e., for $\varphi: \underline{S}(U) \to \Gamma(U,S)$ to be isomorphic for all U) is that S has the following two properties:

1. Each $\underline{S} \mid U$ has the collation property.
2. Each S(V) has no locally zero elements except zero.

It is obvious that $\Gamma(U,F)$ has these properties for any sheaf F. Conversely, if S has these properties the above proposition, applied to U instead of X, shows that $\underline{S}(U) \to \Gamma(U,S \mid U) = \Gamma(U,S)$ is an isomorphism. (Of course, we take Φ to be all closed sets of U.)

The two conditions on S are obviously satisfied by stacks of functions as well as by stacks of sections of vector space bundles.

If a stack contains non-trivial locally zero elements, we cannot apply the above proposition. If, however, we assume Φ is PF, we can dispense with the condition that S has no non-trivial locally zero elements.

Proposition 2:

Let $\underline{S}$ be a stack having the approximate collation property with respect to Φ-coverings. Assume Φ is PF. Then $\underline{S}_\Phi(X) \to \Gamma_\Phi(S)$ is an epimorphism.

Proof:

The proposition follows immediately from the definition of the approximate collation property and the following lemma.

Lemma 2:

Let $\underline{S}$ be any stack and S its associated sheaf. Assume Φ is PF. Let $s \in \Gamma_\Phi(S)$. Then there is a Φ-covering $\{W_\beta\}$ of X and elements $t_\beta \in \underline{S}(W_\beta)$ such that

(a) $\varphi(t_\beta) = s \mid W_\beta$

(b) $\varphi^{W_\beta}_{W_\beta \cap W_\gamma}(t_\beta) = \varphi^{W_\gamma}_{W_\beta \cap W_\gamma}(t_\gamma)$, and

(c) $t_{\beta_o} = 0$ if W_{β_o} is the exceptional set of the Φ-covering.

Remark:

(c) is not needed for the proposition.

Proof:

We follow the method indicated in the remarks preceding the definition of the collation property. Since Φ is PF, the covering $\{N_x\}$ is refined by a Φ-covering $\{U_\alpha\}$. We define $s_\alpha \in \underline{S}(U_\alpha)$ to be $\varphi^{N_x}_{U_\alpha}(s_x)$ for some x with $U_\alpha \subset N_x$. For the exceptional U_{α_o}, we choose $N_x = X - |s|$, so $s_{\alpha_o} = 0$. Let $C \in \Phi$ be such that $U_\alpha \subset C$ for $\alpha \neq \alpha_o$. This C is given by the definition of a Φ-covering.

Shrink $\{U_\alpha\}$ to $\{V_\alpha\}$. Let $x \in X$ be any point. Some neighborhood of x meets only a finite number of U_α's. Take a smaller neighborhood G_x such that

(i) G_x meets $\overline{V}_\alpha$ if and only if $x \in \overline{V}_\alpha$,

(ii) $x \in U_\alpha$ implies $G_x \subset U_\alpha$,

(iii) $x \in V_\alpha$ implies $G_x \subset V_\alpha$, and

(iv) If $x \in \overline{V}_\alpha$ and $x \in \overline{V}_\beta$, then $\varphi^{U_\alpha}_{G_x}(s_\alpha) = \varphi^{U_\beta}_{G_x}(s_\beta)$.

All of these but (iv) are obviously satisfied by all small enough G_x. To satisfy (iv), consider all α such that $x \in \overline{V}_\alpha$. There are only a finite number of such α and each corresponding t_α is defined in a neighborhood of x (namely U_α). All give the same element of the direct limit S_x of $\underline{S}$ (neighborhoods of x). If follows from the properties of direct limits that (iv) is satisfied by all small enough G_x.

Now, for each G_x choose some V_α with $x \in \overline{V}_\alpha$ and define $t_x = \varphi^{U_\alpha}_{G_x}(s_\alpha)$. It follows from (iv) that t_x does not depend on which α is chosen.

I claim that

$$\varphi^{G_x}_{G_x \cap G_y}(t_x) = \varphi^{G_y}_{G_x \cap G_y}(t_y) \text{ for all x and y.}$$

To prove this it is sufficient to consider the case where $G_x \cap G_y \neq \emptyset$. Therefore there is $z \in G_x \cap G_y$. For some α, $z \in V_\alpha$. Therefore G_x and G_y meet $\overline{V}_\alpha$. It now follows from (i) that $x,y \in \overline{V}_\alpha$. Consequently, we may define both t_x and t_y using the same α. The result clearly follows from this.

We can now add another set $G_o = X - C$ to the covering and define $t_o \in \underline{S}(G_o)$ to be zero. This does not spoil the property just proved. In other words,

$$\varphi^{G_o}_{G_o \cap G_x}(t_o) = \varphi^{G_o}_{G_o \cap G_x}(t_x) .$$

To see this, we merely observe that if $G_o \cap G_x \neq \emptyset$, then $G_x \not\subset C$. Since all U_α with $\alpha \neq \alpha_o$ are in C, the only α we can use to define t_x is α_o. Since $s_{\alpha_o} = 0$, we have $t_x = 0$.

Since $C \in \Phi$, the covering $\{G_x, G_o\}$ is refined by a Φ-covering $\{W_\beta\}$. If $\beta \neq \beta_o$, the exceptional index, we choose a $G_x \supset W_\beta$ and define $t_\beta = \varphi^{G_x}_{W_\beta}(t_o)$. If $\beta = \beta_o$, we choose $G_o \supset W_\beta$ and define $t_{\beta_o} = \varphi^{G_o}_{W_{\beta_o}}(t_o) = 0$. The required property of the $\{t_\beta\}$ follows from the corresponding property of the t_x's and t_o.

We now consider stacks of chains and cochains and show that they have the collation property. We start with an indexed collection $\{D_i\}_{i \in I}$ of subsets of X. Assume that for each i there is also given a "base point" $d_i \in D_i$. For example, we can choose a fixed "test space" T with a base point t and let the index set I be the set of all maps $T \to X$. The set D_i will then be defined to be the image of map i or its closure and the point d_i will be defined to be the image of t under the map i. If we let T be the standard n-simplex, then I is the set of all singular n-simplexes of X. If we let T be an ordered set of n + 1 points, then I is the set of all Vietoris n-simplexes of X.

Define $I_U = \{i \in I \mid D_i \subset U\}$. Let M be a module. Define C(U,M) to be the set of functions on I_U with values in M. More generally, if $\overline{M}$ is a protosheaf over X, we let $C(U,\overline{M})$ be the set of functions defined on I_U and such that $f(i) \in \overline{M}_{d_i}$. These $C(U,\overline{M})$ with the obvious restriction maps φ_V^U form a stack $C(*,\overline{M})$.

Lemma 3:

The stack $C(*,\overline{M})$ has the collation property.

Proof:

Let $\{U_\alpha\}$ cover X. Let $t_\alpha \in C(U_\alpha,M)$ be such that

$$\varphi^{U_\alpha}_{U_\alpha \cap U_\beta}(t_\alpha) = \varphi^{U_\beta}_{U_\alpha \cap U_\beta}(t_\beta) .$$

Define $t \in C(X,M)$ as follows:

If D_i is in some U_α, choose one and let $t(i) = t_\alpha(i)$. This is obviously independent of the choice of α.

If D_i is in no U_α, let $t(i) = 0$. Clearly $\varphi^X_{U_\alpha}(t) = t_\alpha$ for all α.

Corollary:

If C_M is the sheaf defined by the stack $C(*,\overline{M})$, then $\Gamma_\Phi(C_M) = C_\Phi(X,M)/\{\text{locally zero elements}\}$ provided Φ is PF.

I do not know if this is true when Φ is not PF.

Remark:

If we start with the Vietoris n-simplexes (defined above), and if M is a module then $C_\Phi(X,M)/\{\text{locally zero elements}\}$ is the classical module of Alexander-Spanier n-cochains with coefficients in M.

If we start with the singular n-simplexes, and if M is a module, then the classical module of singular n-cochains with coefficients in M and supports in Φ is just $C_\Phi(X,M)$. I will show later that the cochain complex of locally zero singular cochains with coefficients in M has zero cohomology. Therefore, factoring it out does not affect the cohomology groups of $C_\Phi(X,M)$. In other words, $C_\Phi(X,M)/\{$locally zero elements$\}$ has the same cohomology as $C_\Phi(X,M)$. To see this, we consider the exact cohomology sequence associated with the short exact sequence $0 \to O(X,M) \to C_\Phi(X,M) \to \frac{C_\Phi(X,M)}{O(X,M)} \to 0$ where $O(X,M)$ is the set of locally zero singular cochains.

The sheaf C_M defined by $C(\ast,\overline{M})$(using any system $\{D_i\}_{i \in I}$ and any protosheaf $\overline{M}$) has another important property.

Proposition 3: The sheaf C_M is fine.

Proof: Let $\{U_\alpha\}$ be a locally finite covering of X. Define a partition of unity $\{w_\alpha\}$ as follows:

For each $x \in X$, choose some α_x such that $x \in U_{\alpha_x}$. Let

$$w_\alpha(x) = \begin{cases} 1 \text{ if } \alpha = \alpha_x \\ 0 \text{ if } \alpha \neq \alpha_x \end{cases}.$$

Define $l'_\alpha : C(U,M) \to C(U,M)$ by : $l'_\alpha(t).(i) = w_\alpha(d_i).t(i)$. These l'_α are stack endomorphisms and so induce $l_\alpha : C_M \to C_M$.

If $x \notin \overline{U}_\alpha$, then x has a neighborhood N_x not meeting $\overline{U}_\alpha$. If $D_i \subset N_x$, then $d_i \notin U_\alpha$. Consequently $w_\alpha(d_i) = 0$ and so $l'_\alpha \mid C(N_x,M) = 0$. Therefore l_α is zero on the stalk of C_M over x. This shows that $|l_\alpha| \subset \overline{U}_\alpha$.

Finally, each $x \in X$ has a neighborhood N_x meeting only a finite member of U_α. The argument just given shows that $l'_\alpha \mid C(N_x,M) = 0$ except for this finite set of α. Since $\Sigma w_\alpha = 1$, it follows that $\Sigma l'_\alpha \mid C(N_x,M) = \text{id}$. Therefore $\Sigma l_\alpha = \text{id}$ in the stalk of C_M over x.

We now come to the stack of chains. Let the system $\{D_i\}_{i \in I}$ be as before. Let $\overline{M}$ be a protosheaf over X. We define $C_\ast(X,\overline{M})$ to be the set of all finite or infinite formal sums $\Sigma m_i\ i$ where $m_i \in \overline{M}_{d_i}$ and $\{D_i \mid m_i \neq 0\}$ forms a locally finite system. We refer to these as locally finite sums. If A is any subset of X, let $C_\ast(A,M)$ be the submodule of $C_\ast(X,M)$ consisting of those sums in which $D_i \subset A$ for all i such that $m_i \neq 0$. Define

$$C_\ast(X,A;M) = \frac{C_\ast(X,M)}{C_\ast(A,M)}.$$

If $B \subset A$, there is an obvious quotient map $C_*(X,B;M) \to C_*(X,A;M)$. Therefore, we can define a stack $\underline{S}$ by letting $\underline{S}(U,M) = C_*(X,X-U;M)$.

Recall that if $\underline{S}$ is a stack and U is open in X, then $\underline{S} \mid U$ is the stack over U defined by those $\underline{S}(V)$ with $V \subset U$.

Proposition 4: Let $\underline{S}$ be the stack defined above. Assume all D_i are connected. Then $\underline{S}$ has the collation property.

Proof: Let U_α be an open covering of X. Let $t_\alpha \in S(U_\alpha,M)$ be elements such that

$$\varphi^{U_\alpha}_{U_\alpha \cap U_\beta}(t_\alpha) = \varphi^{U_\beta}_{U_\alpha \cap U_\beta}(t_\beta).$$

We say that an index i is essential in an open set $V \subset X$ if $D_i \cap V \neq \emptyset$. Two elements $t, t' \in \underline{S}(V)$ are clearly equal if and only if every i essential in V has the same coefficient in t and t'.

I claim that each i has the same coefficient in all t_α for which i is essential in U_α.

To prove this, let i be essential in U_α and U_β. There are two cases to consider:

Case 1: i is essential in $U_\alpha \cap U_\beta$.

Obviously, i has the same coefficient in t_α as in $\varphi^{U_\alpha}_{U_\alpha \cap U_\beta}(t_\alpha)$. A similar result holds for t_β. But,

$$\varphi^{U_\alpha}_{U_\alpha \cap U_\beta}(t_\alpha) = \varphi^{U_\beta}_{U_\alpha \cap U_\beta}(t_\beta).$$

Case 2: The general case.

Since D_i is connected, we can find a "chain" $U_\alpha = U_{\gamma_0}$, $U_{\gamma_1}, \ldots, U_{\gamma_k} = U_\beta$ such that $U_{\gamma_i} \cap U_{\gamma_{i+1}} \cap D_i \neq \emptyset$. But now, case 1 shows that i has the same coefficient in t_{γ_i} and in $t_{\gamma_{i+1}}$.

We now define $t = \Sigma m_i . i$ by choosing m_i to be the coefficient of i in any t_α for which i is essential in t_α. We must show that $\{D_i \mid m_i \neq 0\}$ is locally finite. Let $x \in X$. Then $x \in U_\alpha$ for some α. Therefore, we only have to worry about those D_i which meet U_α. In other words, those i which are essential in U_α. But, these occur in t with the same coefficients as in t_α. Since t_α is a locally finite sum, x has a neighborhood which meets only a finite number of D_i such that $m_i \neq 0$.

This shows that $t \in \underline{S}(X)$. Clearly $\varphi^X_{U_\alpha}(t) = t_\alpha$.

Proposition 5: Assume all D_i are connected.

Let S_M be the sheaf defined by the stack $\underline{S}$. Then $\varphi: \underline{S}(X,M) \to \Gamma(X,S_M)$ is an isomorphism.

Proof: In view of the preceding proposition, it is sufficient to show that $\underline{S}(U,M)$ never has non-trivial locally zero elements. This does not require any condition on the D_i.

Lemma 4: For all U, the only locally zero element of $\underline{S}(U,M)$ is zero itself.

Proof: Let $s = \Sigma m_i . i$ represent a locally zero element of $\underline{S}(U,M)$. Suppose i is essential in U. Then there is a point $x \in U \cap \overline{D}_i$. Since s is locally zero, there is a neighborhood N_x of x such that $\varphi^U_{N_x}(s) = 0$. Since i is essential in N_x, m_i must be zero. Therefore every i essential in U has coefficient zero in s. Thus s is zero in $\underline{S}(U,M)$.

For certain applications, it is useful to have a slightly different definition of the sheaf S_M. Define $C'_*(X,M)$ to be the submodule of $C_*(X,M)$ consisting of those $\Sigma m_i . i$ such that $m_i = 0$ except for a finite number of i. We then define $C'_*(X,A;M)$ and $\underline{S}'(U,M)$ in terms of $C'_*(X,M)$ in exactly the same way that $C_*(X,A;M)$ and $\underline{S}(U,M)$ are defined in terms of $C_*(X,M)$. Obviously $\underline{S}'$ is a substack of $\underline{S}$. Let S'_M be the sheaf it defines.

Lemma 5: The inclusion $\underline{S}' \to \underline{S}$ induces an isomorphism $S'_M \to S_M$.

Proof: Since the inclusion is a monomorphism, it induces a monomorphism of sheaves. Now, let $y \in (S_M)_x$. Choose a representative $s = \Sigma m_i\, i \in S(N_x,M)$ where N_x is some neighborhood of x. Since s is a locally finite sum, some smaller neighborhood U_x meets only a finite number of the D_i for which $m_i \neq 0$. Consequently, $\varphi^{N_x}_{U_x}(s) \in S'(U_x,M)$; but this element also represents y.

Corollary: If M is a module (regarded as a constant sheaf), then $S_M \approx S_K \otimes M$ where K is the ground ring.

Proof: It is obvious that $S'(U,M) \approx S'(U,K) \otimes M$ because

S'(U,K) is a free module (since only finite sums occur).

Another useful property is the following.

Lemma 6: Let $s = \Sigma m_i\, i \in \underline{S}(U,M)$. Then $|s| = \bigcup \bar{D}_i$, the union being taken over those i which are essential in U and such that $m_i \neq 0$.

Proof: Suppose $x \notin |s|$, then there is a neighborhood N_x such that $\varphi^U_{N_x}(s) = 0$. Therefore, no i such that $m_i \neq 0$ can be essential in N_x. This shows that $x \notin \bigcup \bar{D}_i$.

Conversely, suppose $x \notin \bigcup \bar{D}_i$. Then there is a neighborhood N_x disjoint from $\bigcup \bar{D}_i$ because the union of a locally finite system of closed sets is closed. No i with $m_i \neq 0$ is essential in N_x. Therefore, $\varphi^U_{N_x}(s) = 0$, $x \notin |s|$.

To conclude this section, I will prove some additional properties of the sheaves and stacks of singular chains and cochains. The stacks of n-dimensional singular chains and cochains are defined, as indicated above, by taking the set of indices I to be the set of singular n-simplexes of X. If we take M to be a module, we can define boundary and coboundary operators by the usual formulas. In this way we arrive at chain and cochain complexes of stacks and sheaves. We will need the following lemma of singular homology theory:

Lemma 7: Let $\mathfrak{U} = \{U_\alpha\}$ be a collection of subsets of X whose interiors cover X. Let $\underline{S}'_{\mathfrak{U}}(X) = C'_{\mathfrak{U}}(X,K)$ be the set of finite singular chains (with coefficients in the ground ring K) which contain only simplexes i for which Im(i) is contained in some set of $\mathfrak{U}$. Let $i: \underline{S}'_{\mathfrak{U}}(X) \to \underline{S}'(X)$ be the inclusion map.

Then, there is a chain map $j: \underline{S}'(X) \to \underline{S}'_{\mathfrak{U}}(X)$ and a homotopy $D: \underline{S}'(X) \to \underline{S}'(X)$ such that

(a) $j\, i = id$,

(b) $id. - i\, j = \partial D + D\partial$, and

(c) j and D (and i) preserve supports, i.e. $|js| \subset |s|$ and $|Ds| \subset |s|$.

For a proof, we refer the reader to Eilenberg and Steenrod, "Foundations of Algebraic Topology", pp. 207-8.

It is now quite easy to prove the following proposition which was mentioned earlier.

Proposition 6: Let M be any module and O(X,M) the set of all locally zero singular cochains of X with coefficients in M. Then H(O(X,M)) = O.

Proof: Let $v \in O(X,M)$ be a cocycle. Since v is locally zero, there is a covering $\mathfrak{U}$ of X such that $\varphi^X_{U_\alpha}(v) = 0$ for all α. Clearly, v annihilates $\underline{S}_{\mathfrak{U}}(X)$. Let i, j, D be the maps defined in lemma 7. Let $i^\#$, $j^\#$, $D^\#$ be the maps induced on Hom($\underline{S}_{\mathfrak{U}}(X)$,M) and Hom($\underline{S}'(X)$,M) = C(X,M). Then, $i^\#(v) = 0$. Therefore $v = (id - ij)^\# v = (\partial D + D\partial)^\# v = (\delta D^\# + D^\# \delta)v = \delta D^\# v$. This shows that every cocycle of O(X,M) cobounds.

We now state and prove an important property of the sheaf S_K of singular chains.

Definition: Let F be a chain (or cochain) complex of sheaves. We say F is homotopically fine if, for every locally finite covering $\{U_\alpha\}$ of X, we can find endomorphisms l_α and D of F such that

(a) $|l_\alpha| \subset \overline{U}_\alpha$, and

(b) $\Sigma l_\alpha = id - \partial D - D\partial$.

Remark: This is equivalent to saying that F is $\mathcal{C}$-fine where $\mathcal{C}$ is the class of endomorphisms of F which are chain homotopic to the identity. Note that the l_α are not required to be chain maps.

Proposition 7: The sheaf S_K of singular chains is homotopically fine.

Proof: Define a stack $\underline{S}'_{\mathfrak{U}}(U,K)$ in terms of $C'_{\mathfrak{U}}(X,K)$ in exactly the same way that $\underline{S}'(U,K)$ is defined in terms of C'(X,K). Since i, j, and D preserve supports, they induce maps of the modules occurring in these stacks.

Define l'_α: $\underline{S}'_{\mathfrak{U}}(U) \to \underline{S}'_{\mathfrak{U}}(U)$ as follows: Let i be a singular simplex such that T_i = Im i is contained in some U_{α_0}. Choose one such α_0 and let $l'_{\alpha_0}(i) = i$ and $l'_\alpha(i) = 0$ for $\alpha \neq \alpha_0$.

Then, we have $|l'_\alpha(s)| \subset U_\alpha$ for all s, and $\Sigma l'_\alpha = id$.

Now, define l''_α: $\underline{S}'(U) \to \underline{S}'(U)$ by $l''_\alpha = i\, l'_\alpha\, j$. Then, since i and j preserve supports, $|l''(s)| \subset U_\alpha$ for all s. Also, $\Sigma l''_\alpha = i(\Sigma l'_\alpha)j = i\,j = id - \partial D - D\partial$.

Finally, the $1''_\alpha$ and D induce endomorphisms of S_K. If $x \notin \overline{U}_\alpha$, there is a neighborhood N_x of x disjoint from U_α. If $s \in \underline{S}(N_x)$, then, since $|1''_\alpha(s)| \subset U_\alpha$, we must have $1''_\alpha(s) = 0$ (because $|1''_\alpha(s)| \subset N_x$ also and so $|1''_\alpha(s)| = \emptyset$). Therefore 1_α is zero on the stalk over x. This shows that $|1_\alpha| \subset \overline{U}_\alpha$.

For any $x \in X$, there is a neighborhood N_x meeting only a finite number of U_α. Therefore all but a finite number of $1''_\alpha$ are zero on $\underline{S}(N_x)$. Since $\Sigma 1''_\alpha = \text{id} - \partial D - D\partial$, it follows that the 1_α have the same property.

Corollary: S_M is homotopically fine for all modules M.

Proof: Since $S_M \approx S_K \otimes M$, it is sufficient to find the 1_α and D for S_K and then use $1_\alpha \otimes 1$ and $D \otimes 1$ as the corresponding maps for S_M.

Our last lemma shows that singular homology based on finite chains is the same as singular homology with compact supports.

Lemma 8: Let Φ be the set of all compact subsets of X. Let $\underline{S}$, $\underline{S}'$ be the stacks of singular chains defined above. Then $\underline{S}'(X,M) = \underline{S}_\Phi(X,M)$ for all M.

Proof: The D_i are all compact. Therefore all finite unions of D_i are compact, and so, $\underline{S}'(X,M) \subset \underline{S}_\Phi(X,M)$. Conversely, let $s \in \underline{S}_\Phi(X,M)$. Then $|s|$ is the union of a locally finite system of D_i. To show $s \in \underline{S}'(X,M)$, it is sufficient to show that the union of a locally finite system of compact sets is compact if and only if the system is finite. This follows from the fact that one can cover the union by a collection of open sets each meeting only a finite number of sets of the system. If the union is compact, it can be covered by a finite number of such sets.

VII. MISCELLANEOUS RESULTS

<u>INDUCED SHEAVES</u> Let X and Y be spaces and $f:X \to Y$ any map. Let F be a sheaf over X and G a sheaf over Y.

<u>Definition</u>: A map $\varphi:G \dashrightarrow F$ over f is a collection of maps $\varphi_x:G_{f_x} \to F_x$, one for each $x \in X$ satisfying the following condition:

If U is open in Y and $t \in \Gamma(U,G)$, define $s:f^{-1}(U) \to F$ by $s(x) = \varphi_x t(f(x))$. Then s is continuous (and therefore $s \in \Gamma(f^{-1}(U),F)$). (We write $s = \varphi_{\#}(t)$.)

Note that the map φ is not given by a set theoretical mapping $G \to F$. For this reason we write $\varphi:G \dashrightarrow F$ with a dotted line in the arrow. Almost by definition, a map $\varphi:G \dashrightarrow F$ induces a map $\varphi_{\#}:\Gamma(U,G) \to \Gamma(f^{-1}(U),F)$ for all open U in Y. Obviously, if $s \in \Gamma(U,G)$, then $|\varphi_{\#}(s)| = f^{-1}(|s|)$. Suppose Φ is a family of supports in X, Ψ a family of supports in Y and $A \in \Psi$ implies $f^{-1}(A) \in \Phi$. Then φ induces a map $\varphi_{\#}:\Gamma_\Psi(G) \to \Gamma_\Phi(F)$.

An example of such a mapping is given by the canonical mapping from a sheaf G to the induced sheaf $f^{-1}(G)$ defined as follows.

<u>Definition</u>: Let $f:X \to Y$ be any map. Let G be a sheaf over Y with projection p. Define $f^{-1}(G) = \{(x,g) \in X \times G \mid f(x) = p(g)\}$. Define $p:f^{-1}(G) \to X$ by $p'(x,y) = x$. For each $x \in X$, define $i_x:G_{f(x)} \to f^{-1}(G)_x$ by $i_x(g) = (x,g)$. Then i_x is 1-1 onto. We make $f^{-1}(G)_x$ a K-module by transporting the K-module structure of $G_{f(x)}$ to $f^{-1}(G)_x$ by means of i_x.

<u>Lemma 1</u>: $f^{-1}(G)$ is a sheaf over X and i is a map $i:G \dashrightarrow f^{-1}(G)$ in the sense defined above.

The verification of this lemma is left to the reader.

Now, let $f:X \to Y$ be some fixed map: Let $\mathcal{A}$ be the category of sheaves over Y and $\mathcal{B}$ the category of sheaves over X. Define an $\mathcal{A}\mathcal{B}$-map to be a map φ in the sense defined above. It is easy to check that these maps have all the properties required of $\mathcal{AB}$-maps so that we can speak of universal $\mathcal{AB}$-maps.

<u>Proposition 1</u>: The maps $i: G \dashrightarrow f^{-1}(G)$ are universal $\mathcal{AB}$-maps.

<u>Proof</u>: Let $\varphi: G \dashrightarrow F$ be an $\mathcal{AB}$-map. Define $h: f^{-1}(G) \to F$ by letting $h \mid f^{-1}(G)_x$ be the composition

$$f^{-1}(G)_x \xrightarrow[i_x^{-1}]{\approx} G_{f(x)} \xrightarrow[\varphi_x]{} F_x .$$

Obviously h is a map of protosheaves and $\varphi = h \circ i$. To show that h is continuous, let $y \in f^{-1}(G)_x$. Let s be a local section of G through $i_x^{-1}(y)$. Then $i_{\#}(s)$ is a local section through y and $\varphi_{\#}(s)$ is a local section through h(y). But, $h\, i_{\#}(s) = \varphi_{\#}(s)$. Therefore h preserves local sections and so is continuous.

As a consequence of this proposition, we need never consider maps of the type $\varphi: G \dashrightarrow F$ other than i. They can all be replaced by maps $f^{-1}(G) \to F$.

Since $f^{-1}(G)$ can be defined by a universal property, it is obviously a functor in G. In fact, f^{-1} is an exact functor since each stalk of $f^{-1}(G)$ is isomorphic to a stalk of G under i.

Now, let Φ be a family of supports in X and Ψ a family of supports in Y such that $A \in \Psi$ implies $f^{-1}(A) \in \Phi$. Consider the composition of the cohomology functor $H_\Phi(X, \ )$ with the functor f^{-1}, i.e., the functor sending G to $H_\Phi(X, f^{-1}(G))$. Since f^{-1} is exact, this composition is an exact δ-functor. We can define an augmentation of it over $\Gamma_\Psi(G)$ to be the composition $\Gamma_\Psi(G) \xrightarrow{i_{\#}} \Gamma_\Phi(f^{-1}(G)) \xrightarrow{\epsilon} H^0_\Phi(X, f^{-1}(G))$.

Since the cohomology functor $H_\Psi(Y, \ )$ is a universal augmented δ-functor over $\Gamma_\Psi(G)$, there is a unique map of augmented δ-functors $H_\Psi(Y,G) \to H_\Phi(X, f^{-1}(G))$. We call this the map of cohomology induced by $f: X \to Y$. If $\varphi: G \dashrightarrow F$ is a map over F, then we factor φ into $G \overset{i}{\dashrightarrow} f^{-1}(G) \xrightarrow{h} F$ and define $\varphi_*\ H_\Psi(Y,G) \to H_\Phi(X,F)$ to be the composition $H_\Psi(Y,G) \to H_\Phi(X, f^{-1}(G)) \xrightarrow{h_*} H_\Phi(X,F)$.

Note that we can take X = Y and f = id. The above construction then gives us a unique map of augmented δ-functors $H_\Psi(X,G) \to H_\Phi(X,G)$ whenever $\Phi \supset \Psi$.

SUBSPACES

Suppose $A \subset X$ is locally closed. Let Φ be a family of supports in X. Define $\Phi_A = \{B \in \Phi \mid B \subset A\}$. We want to try and express $H_{\Phi_A}(A,\ \ )$ in terms of $H_\Phi(X,\ \ )$. In general, the families Φ and Φ_A do not satisfy the conditions imposed on Ψ and Φ in the case of a map $f:X \to Y$. To satisfy this condition, it would be necessary to take $\{A \cap B \mid B \in \Phi\}$ in place of Φ_A. We must therefore use a different method. This is based on the following lemma.

Lemma 2: If A is locally closed and F is a sheaf over X, then $\Gamma_\Phi(F_A) \approx \Gamma_{\Phi_A}(F \mid A)$ naturally.

Proof: Define $\Gamma_\Phi(F_A) \to \Gamma_{\Phi_A}(F \mid A)$ by restricting sections to A. Define $\Gamma_{\Phi_A}(F \mid A) \to \Gamma_\Phi(F_A)$ by prolonging sections by zero. These maps are obviously inverses. They are easily seen to be well defined.

Corollary: If A is locally closed and G is a sheaf over A, then $\Gamma_{\Phi_A}(G) \approx \Gamma_\Phi(G^X)$ naturally.

Proof: $G^X \mid A = G$ and $(G^X)_A = G^X$.

Now, $*^X$ is an exact functor. Therefore the composition of this with $H_\Phi(X,\ \ )$ is an exact δ-functor. This composition sends G to $H_\Phi(X,G^X)$. In dimension zero, $H^0_\Phi(X,G^X) = \Gamma_\Phi(G^X) \approx \Gamma_{\Phi_A}(G)$ naturally. Therefore, $H_\Phi(X,*^X)$ is an augmented δ-functor over Γ_{Φ_A}. By the universal property of $H_{\Phi_A}(A,\ \ )$, there is a unique map of augmented δ-functors $H_{\Phi_A}(A,\ \ ) \to H_\Phi(X,*^X)$.

Note that $H_\Phi(X,*^X)$ satisfies all conditions for a cohomological δ-functor over Γ_{Φ_A} with the exception of the effaceability condition.

Proposition 2: This map $H_{\Phi_A}(A,\ \ ) \to H_\Phi(X,*^X)$ is a natural equivalence of augmented δ-functors if either

(1) A is closed

or (2) A is locally closed and Φ is PF.

Proof: It is sufficient to show $H^p_\Phi(X,*^X)$ is effaceable for $p > 0$ under either of the given conditions. This follows immediately from the following lemmas.

Lemma 3: If A is closed, F is a sheaf over X, and G a sheaf

over A, then Hom $(F \mid A, G) \approx$ Hom (F, G^X) naturally and corresponding elements have the same support. (This gives us another example of adjoint functors.)

Proof: Define Hom $(F, G^X) \approx$ Hom $(F \mid A, G)$ by restriction; i.e., $f \to f \mid A$ if $f: F \to G^X$. Obviously, $|f \mid A| = |f|$ since A is closed and since f must be zero on stalks over points $x \notin A$.

Define Hom $(F \mid A, G) \to$ Hom (F, G^X) by prolongation by zero. This is well defined if A is closed. It is clearly the inverse of the restriction map.

Corollary: If A is closed and G is an injective sheaf over A, then G^X is injective.

Proof: We must show that Hom $(\ , G^X)$ is an exact functor. But it is naturally equivalent to Hom $(\ \mid A, G)$ which is a composition of the exact functors $\ast \mid A$ and Hom $(\ , G)$.

Lemma 4: If A is locally closed in X and G is a fine sheaf over A, then G^X is also fine.

Proof: Let $\{U_\alpha\}$ be a locally finite covering of X. Then $\{U_\alpha \cap A\}$ is a locally finite covering of A. Let $1_\alpha': G \to G$ be endomorphisms such that $|1_\alpha'| \subset \overline{U_\alpha \cap A}$ and $\Sigma 1_\alpha' = \text{id}$.

The $1_\alpha'$ induce maps $1_\alpha: G^X \to G^X$ (since prolongation by zero is a functor). Obviously $|1_\alpha| \subset \overline{U}_\alpha$ and $\Sigma 1_\alpha = \text{id}$.

To prove the proposition, let G be injective. Then G is also fine. In case (1), G^X is also injective. In case (2) G^X is fine. Therefore, in either case $H^p_\Phi(X, G^X) = 0$ for $p \neq 0$.

Corollary 1: Let F be any sheaf over X.
If either (1) A is closed in X
or (2) A is locally closed and Φ is PF,
then $H_\Phi(X, F_A) \approx H_{\Phi_A}(A, F \mid A)$.
This is a natural equivalence of augmented δ-functors.

Proof: Let $G = F \mid A$. Then $G^X = F_A$ by definition.

Suppose now that A is closed in X. Let $U = X - A$. Then we have the exact sequence $0 \to F_U \to F \to F_A \to 0$. Take the exact cohomology sequence of this and apply corollary 1. This gives the exact sequence $\ldots \to H^p_\Phi(X, F_U) \to H^p_\Phi(X, F) \to H^p_{\Phi_A}(A, F \mid A) \xrightarrow{\delta} \ldots$

If, in addition Φ is PF, we may apply the corollary to replace $H^p_\Phi(X, F_U)$ by $H^p_{\Phi_U}(U, F \mid U)$. The resulting sequence is analagous to the cohomology sequence of a pair.

If A is closed in X, the families Φ and Φ_A satisfy the conditions imposed on Φ and Ψ in the consideration of a map from one space to another. Let $j: A \to X$ be the inclusion. Obviously $j^{-1}(F) = F \mid A$ for any sheaf F over X. Therefore, there is a unique map of augmented δ-functors $j^*: H_\Phi(X,F) \to H\Phi_A(A,F \mid A)$. But, such a map occurs in the above exact sequence, the map being induced by $F \to F_A$. Consequently, the map $H_\Phi(X,F) \to H\Phi_A(A,F \mid A)$ occurring in the exact sequence is the map j^* of cohomology induced by the inclusion $A \to X$.

Suppose finally that $F = G^X$ where G is a sheaf over A. Then $j^*: H_\Phi(X,G^X) \to H\Phi_A(A,G)$. This is a map of cohomological δ-functors and so must give the canonical isomorphism. Therefore, if A is closed in X, the canonical isomorphism $H\Phi_A(A,G) \approx H_\Phi(X,G^X)$ is given by $(j^*)^{-1}$.

We will now try to calculate $H_\Psi(A,F \mid A)$ where Ψ is any family of supports on A, not necessarily of the form Φ_A. Since restriction to A is an exact functor, $H_\Psi(A,F \mid A)$ is an exact δ-functor, augmented over $\Gamma_\Psi(F \mid A)$. We will show that $H^i_\Psi(A,F \mid A)$ is effaceable (as a functor in F not $F \mid A$) for $i > 0$ under certain conditions.

<u>Lemma 5</u>: Assume either (1) A is open

<u>or</u> (2) $\Psi = \Phi_A$ for some family Φ on X and Ψ and Φ are both PF.

Then $H^p_\Psi(A,F \mid A)$ is effaceable (as a functor in F) for all $p \neq 0$.

<u>Proof</u>: This follows immediately from the fact that if A is open in X and F is flabby, then $F \mid A$ is flabby (lemma IV. 1, p. 59).

<u>Lemma 6</u>: If F is Φ-soft$_3$, then $F \mid A$ is Φ_A-soft$_3$.

<u>Proof</u>: Let $B \in \Phi_A$ and $s \in \Gamma(B,F \mid A)$. Since $\Gamma(B,F \mid A) = \Gamma(B,F)$, we can extend s to X and then restrict this extension to B.

The next lemma gives us a sufficient condition for Φ_A to be PF.

<u>Lemma 7</u>: If A is locally closed and Φ is PF, then Φ_A is PF.

<u>Proof</u>: The sets of Φ_A, being sets of Φ, are clearly paracompact. Since A is locally closed, $A = U \cap Y$ with U open

and Y closed. Let $B \in \Phi_A$ and let C be a neighborhood of B in Φ. Then $U \cap \text{Int } C$ is a neighborhood of B in C. Since C is paracompact, it is normal. Therefore there is a closed subset D of C such that $B \subset \text{Int } D$ and $D \subset U \cap \text{Int } C$. Here Int D means the interior of D as a subset of C, but D is contained in $U \cap \text{Int } C$ which is open in X. Therefore, Int D is also the interior of D as a subset of X.

Now, $D \cap A$ is a neighborhood of B in A. But, $D \cap A = D \cap U \cap Y = D \cap Y$ is closed in C. Therefore $D \cap A \in \Phi$ and so $D \cap A \in \Phi_A$.

Proposition 3: Let $A \subset X$. Let Ψ be a family of supports on A.

Assume either

(1) A is open

or (2) A is locally closed and

$\Psi = \Phi_A$ where Φ is a PF family of supports on X.

Then $H_\Psi(A,F \mid A)$, considered as a functor in F (not $F \mid A$), is a cohomological δ-functor over $\Gamma_\Psi(F \mid A)$.

Proof: This is an immediate consequence of the above lemmas.

Corollary 2: $H^i_\Psi(A,F \mid A)$ is the ith right derived functor of $\Gamma_\Psi(F \mid A)$ (by the uniqueness of cohomological δ-functors).

Corollary 3: Let A be open in X. Suppose $\Phi = \Phi_A$. Then $H_\Phi(X,F) \approx H_{\Phi_A}(A,F \mid A)$ for all sheaves F over X. This is a natural equivalence of δ-functors.

Proof: It is sufficient to show that $\Gamma_\Phi(F) \approx \Gamma_{\Phi_A}(F \mid A)$ naturally for all F and then use the uniqueness of cohomological δ-functors.

Let $\Gamma_\Phi(F) \to \Gamma_{\Phi_A}(F \mid A)$ be the restriction map. Let $\Gamma_{\Phi_A}(F \mid A) \to \Gamma_\Phi(F)$ be prolongation by zero. This is well defined because if $s \in \Gamma_{\Phi_A}(F \mid A)$, then the prolongation of s agrees with s on A and is zero over $X - |s|$. Therefore it is obtained by amalgamating two sections, one over A and one over $X - |s|$, which agree on $A \cap (X - |s|)$. This shows that the prolongation is continuous.

The restriction and prolongation maps are obviously inverses.

Corollary 4: Let A be any subset of X. Let Φ be a family of supports in X such that $\Phi = \Phi_A$ and such that every set of Φ has a neighborhood in Φ. Then $H_\Phi(X,F) \approx H_{\Phi_A}(A,F\,|\,A)$ for all F. This is a natural equivalence of δ-functors.

Proof: Let U be the union of the interiors of all sets of Φ. Then U is open in X, $U \subset A$, and $\Phi_U = \Phi$. By the preceding corollary, $H_\Phi(X,F) \approx H_{\Phi_U}(U,F\,|\,U)$. Another application of the same collary with A in place of X shows $H_{\Phi_A}(A,F\,|\,A) \approx H_{\Phi_U}(U,F\,|\,U)$. This gives the desired result.

Corollary 5: Let $A \subset X$. Assume $\Phi = \Phi_A$.

Assume either (1) A is open in X

or (2) A is locally closed and every set in Φ has a neighborhood in Φ.

Then, $H_\Phi(X,F) \approx H_\Phi(X,F_A)$ for all sheaves F over X. This is a natural equivalence of δ-functors.

Proof: Using the preceding corollary, we get $H_\Phi(X,F) \approx H_{\Phi_A}(X,F\,|\,A) \approx H_\Phi(X,F_A)$ since $F_A\,|\,A = F\,|\,A$. The fact that A is locally closed is needed only to show the existence of F_A.

CONTINUITY PROPERTIES

If $\{T_\alpha\}$ is a direct system of covarient linear functors, we define $T = \varinjlim T_\alpha$ to be the composition of the functor $\{T_\alpha\}$ from objects to direct systems and the functor $\varinjlim$ from direct systems to objects. Obviously, $T(A) = \varinjlim T_\alpha(A)$ for all A.

Lemma 8: Let $T = \varinjlim T_\alpha$. Then $R^*T = \varinjlim R^*T_\alpha$ as an augmented δ-functor.

Proof: If A is any object, let C be an injective resolution of A. Then $R^*T(A) = H(T(C)) = H(\varinjlim T_\alpha(C)) = \varinjlim R^*T_\alpha(C)$. If A is a short exact sequence, let C be a resolution composed of short exact sequences. The same argument then shows that $R^*T = \varinjlim R^*T_\alpha$ as a δ-functor. The augmentation is trivially preserved.

Proposition 4: Suppose Φ is PF. Then we have the natural equivalence of augmented δ-functors $H_\Phi(X,F) \approx \varinjlim H(X,F_U)$ for all sheaves F over X, the limit being taken over all open U such that $\overline{U} \in \Phi$ and the maps of the direct system being given by the inclusions $F_V \to F_U$ if $V \subset U$.

(Note: If no supports are specified (as in $H(X,F_U)$, the family of supports should be taken to be the set of all closed subsets of X).

Proof: If $\overline{U} \in \Phi$, then $\Phi_U = \{$all closed subsets of X contained in U$\}$. By corollaries 1 and 5 of the last section, $H(X,F_U) \approx H_{\Phi_U}(U,F \mid U) \approx H_{\Phi_U}(X,F)$ for all F. Consequently, the $H^i(X,F_U)$ are right derived functors of Γ_{Φ_U}. Therefore, the result will follow from the lemma if we show that $\Gamma_\Phi(F) = \varinjlim \Gamma_{\Phi_U}(F)$ for all F. But $\Gamma_{\Phi_U}(F) \subset \Gamma_\Phi(F)$ for all F, and, if $s \in \Gamma_\Phi(F)$, then $s \in \Gamma_{\Phi_U}(F)$ where $U \supset |s|$ is an open set such that $\overline{U} \in \Phi$. Such a U exists by the hypothesis.

Suppose now that A is some locally closed subset of X.

Let $\{U_\alpha\}$ be a collection of open sets, all containing A, and directed by inclusion. Let Ψ_α be a family of supports in U_α such that $U_\beta \subset U_\alpha$ and $B \in \Psi_\alpha$ imply $B \cap U_\beta \in \Psi_\beta$. Then, if $U_\beta \subset U_\alpha$, there is a restriction map $\Gamma_{\Psi_\alpha}(F \mid U_\alpha) \to \Gamma_{\Psi_\beta}(F \mid U_\beta)$ for all sheaves F over X. The resulting maps of cohomological δ-functors are the maps of cohomology $H_{\Psi_\alpha}(U_\alpha,F \mid U_\alpha) \to H_{\Psi_\beta}(U_\beta,F \mid U_\beta)$ induced by the inclusions $U_\beta \to U_\alpha$. Since the set of U_α was assumed directed, these cohomology groups form a direct system.

Proposition 5: Let A be locally closed in X. Let Φ be a PF family of supports in X. Let the U_α and Ψ_α be as above. Assume $\Gamma_{\Phi_A}(F \mid A) \approx \varinjlim \Gamma_{\Psi_\alpha}(F \mid U_\alpha)$ naturally for all sheaves F over X.

Then, $\varinjlim H_{\Psi_\alpha}(U_\alpha,F \mid U_\alpha) \approx H_{\Phi_A}(A,F \mid A)$ for all F. This is a natural equivalence of augmented δ-functors.

Proof: By corollary 2 of the last section, $H_{\Psi_\alpha}(U,F \mid U_\alpha)$ and $H_{\Phi_A}(A,F \mid A)$ are the right derived functors of $\Gamma_{\Psi_\alpha}(F \mid U_\alpha)$ and $\Gamma_{\Phi_A}(A \mid F)$, respectively. Therefore the result follows immediately from the lemma.

Example 1: Let X be paracompact and Φ the set of all closed subsets of X. Let A be closed in X. Let $\{U_\alpha\}$ be all open sets containing A. Let Ψ_α be all (relatively) closed subsets of U_α.

Then $H(A,F \mid A) = \varinjlim_{\substack{U \text{open} \\ U \supset A}} H(U,F \mid U)$.

Proof: We have only to verify that $\Gamma_{\Phi_A}(F \mid A) \approx \varinjlim \Gamma_{\Psi_\alpha}(F \mid U_\alpha)$ i.e., that $\Gamma(A,F) = \varinjlim_{\substack{U \text{ open} \\ U \supset A}} \Gamma(U,F)$. This follows from the fact that two sections over U which agree on A must agree on a neighborhood of A, and the fact that any section over A extends to a neighborhood of A.

For the next example, we need a preliminary lemma.

Lemma 9: Let S and T be left exact functors on the category of sheaves over X. Let $f:S \to T$ be a natural transformation such that $f_I:S(I) \to T(I)$ is an isomorphism when I is injective. Then f is a natural equivalence.

Proof: Let A be any object and $0 \to A \to I^o \to I^1 \to \ldots$ an injective resolution. Then

$$\begin{array}{ccccccc} 0 \to & S(A) & \to & S(I^o) & \to & S(I^1) \\ & \downarrow f_A & & \downarrow \approx & & \downarrow \approx \\ 0 \to & T(A) & \to & T(I^o) & \to & T(I^1) \end{array}$$

The 5-lemma shows that f_A is an isomorphism.

Example 2: Let X be locally compact. Let Φ be the set of all compact subsets of X. Let A be closed in X. Let $\{U_\alpha\}$ be a directed system of open sets containing A such that $\bigcap \overline{U}_\alpha = A$. (It is not necessary to take all open sets containing A.) Let Ψ_α be the collection of sets $B \cap U_\alpha$ where B runs over all compact sets of X.

Then $H_{\Phi_A}(A,F \mid A) = \varinjlim H_{\Psi_\alpha}(U_\alpha,F \mid U_\alpha)$.

Proof: Φ is clearly PF. We have only to verify that $\Gamma_{\Phi_A}(F \mid A) = \varinjlim \Gamma_{\Phi_\alpha}(F \mid U_\alpha)$. There is an obvious map $\varinjlim \Gamma_{\Psi_\alpha}(F \mid U_\alpha) \to \Gamma_{\Phi_\alpha}(F \mid A)$. We have only to show that this map is an isomorphism when F is injective. If F is injective, it is Φ-soft. Therefore, any section $s \in \Gamma_{\Phi_A}(F \mid A)$ extends to a section $s' \in \Gamma_\Phi(F)$. The restrictions $s' \mid U_\alpha$ are all preimages of s. Therefore, the map is an epimorphism.

Suppose now that $s \in \Gamma_{\Psi_\alpha}(F \mid U_\alpha)$ and $s \mid A = 0$. Let $|s| = B \cap U_\alpha$ where B is compact. Since $s \mid A = 0$, $B \cap A = \emptyset$. But, $\bigcap \overline{U}_\alpha = A$ and so, $\bigcap \overline{U}_\alpha \cap B = \emptyset$. Since B is compact and $\{U_\alpha\}$ is directed, there is a β with $\overline{U}_\beta \cap B = \emptyset$. Therefore $s \mid U_\beta = 0$. This shows that the map is a monomorphism.

ALEXANDER-SPANIER COHOMOLOGY

Let M be a module and $\check{C}_M$ the sheaf of germs of Alexander-Spanier cochains with coefficients in M (see p. 28). Then $0 \to M \xrightarrow{\epsilon} \check{C}^o_M \to \check{C}^1_M \to \ldots$ is exact. Also, the $\check{C}^i_M$ are fine.

Therefore, if Φ is PF, $\check{C}_M$ is a Φ-resolution of M. Thus $H(\Gamma_\Phi(\check{C}_M)) \approx H_\Phi(X,M)$ naturally. But, since Φ is PF, $\Gamma_\Phi(\check{C}_M)$ is just the classical module of Alexander-Spanier cochains with coefficients in M and supports in Φ. Therefore the classical Alexander-Spanier cohomology groups agree with $H_\Phi(X,M)$ when Φ is PF.

We can define a pairing $\check{C}_M \otimes \check{C}_N \to \check{C}_{M \otimes N}$ by the classical Alexander-Čech-Whitney formula. We define first a map of stacks. If U is open in X, $\Psi^p(U,M) \otimes \Psi_q(U,N) \to \Psi^{p+q}(U,M \otimes N)$ is defined by taking $f \in \Psi^p(U,M)$ g $\in \Psi^q(U,N)$ and defining $f.g \in \Psi^{p+q}(U,M \otimes N)$ by $(f.g).(x_o,\ldots,x_{p+q}) = f(x_o,\ldots,x_p) \otimes g(x_p,\ldots,x_{p+q})$.

This map is easily seen to be an augmentation preserving cochain map. Therefore, $\check{C}_M \otimes \check{C}_N \to \check{C}_{M \otimes N}$ is also an augmentation preserving cochain map. If $\Phi = \Phi_1 \quad \Phi_2$ with Φ_1, Φ_2 (and therefore Φ) PF, this map induces the classical cup product $H^p_{\Phi_1}(X,M) \otimes H^q_{\Phi_2}(X,N) \to H^{p+q}_\Phi(X,M \otimes N)$. We want to show that this agrees with the cup product given by the theory of sheaves.

Since $\check{C}_M$ is algebraically split, there is an augmentation preserving cochain map f into the canonical flabby resolution $PQ^*(M)$. The same holds for N. These maps induce the canonical isomorphism of the classical Alexander-Spanier groups $H(\Gamma_\Phi(\check{C}_M))$ with the groups $H_\Phi(X,M)$.

Let I be an injective resolution of $M \otimes N$. Choose a map $\check{C}_{M \otimes N} \to I$. This also induces the canonical isomorphism $H(\Gamma_\Phi(\check{C}_{M \otimes N})) \approx H_\Phi(X,M \otimes N)$. Choose a map $PQ^*(M) \otimes PQ^*(N) \to I$. This induces the cup product used in the theory of sheaves.

Now, $\check{C}_M \otimes \check{C}_N$ is acyclic over $M \otimes N$ since it is the tensor product of two algebraically split resolutions. Therefore, the following diagram commutes up to cochain homotopy

$$\begin{array}{ccc} \check{C}_M \otimes \check{C}_N & \to & \check{C}\, M \otimes N \\ \downarrow & & \downarrow \\ PQ^*(M) \otimes PQ^*(N) & \to & I \end{array}$$

This shows that the two cup products agree.

<u>SINGULAR COHOMOLOGY</u>

Let M be a module and C_M the sheaf of germs of singular

cochains with coefficients in M. The C_M^i are again fine but

$$0 \to M \overset{\epsilon}{\to} C_M^o \to C_M^1 \to \dots$$

will not be exact in general. We recall that if this sequence is exact, X is said to be HLC. If X is HLC, then we obviously have $H(\Gamma_\Phi(C_M)) \approx H_\Phi(X,M)$ whenever Φ is PF.

This isomorphism can be realized by an explicit map from the Alexander-Spanier cohomology to the singular cohomology. This map is always defined even if X is not HLC.

Let $\Delta^n = \{(t_o,\dots,t_n) \mid t_i \geq 0, \Sigma t_i = 1\}$ be the standard n-simplex. Let $v_i \in \Delta^n$ be the point $(0,\dots,1,0,\dots,0)$ with $t_i = 1$. Define a map $j\colon \Psi^n(U,M) \to C^n(U,M)$ as follows:

Let $w \in \Psi^n(U,M)$.

If $h\colon \Delta^n \to U$ is any singular simplex, define $jw.(h) = w(h(v_o), h(v_1),\dots, h(v_n))$.

It is easy to see that j is an augmentation preserving cochain map and preserves cup products. Therefore the same is true of the induced map $j\colon \check{C}_M \to C_M$ and of the map of Alexander-Spanier cohomology. This map of cohomology will be an isomorphism if Φ is PF and X is HLC. Therefore, when Φ is PF and X is HLC, the singular, Alexander-Spanier, and sheaf cohomology groups with coefficients in a module are all naturally isomorphic. These isomorphisms preserve cup products.

In the next chapter, we shall see that the same is true for the Čech groups.

VIII. ČECH COHOMOLOGY

I will first define the Čech cohomology groups of X with coefficients in a stack. These groups are defined for all families Φ and all directed systems of coverings. There are two ways to define these groups, depending on how we define supports in the cochain modules. The natural definition of supports make use of the concept of the support of an element of a stack, as defined in the previous section. This leads to a system of cohomology groups $\check{H}^*_\Phi(X,\underline{S})$ which are augmented but do not form a δ-functor. To remedy this, we use a cruder definition of supports. This gives a system $\check{H}^*_\Phi(X,\underline{S})'$ which is an exact δ-functor but is not augmented in general. The two definitions agree if Φ is the collection of all closed subsets of X.

We then define Čech cohomology with coefficients in a sheaf F by taking the stack $\underline{S}$ to be $\Gamma(\ ,F)$. This gives us two systems of groups $\check{H}^*_\Phi(X,F)$ and $\check{H}^*_\Phi(X,F)'$. The functors $H^p_\Phi(X,F)$ are effaceable for $p > 0$, and $\epsilon: \Gamma_\Phi(F) \to H^0_\Phi(X,F)$ is an isomorphism. Also, if Φ and the directed system of coverings satisfy certain simple conditions, there is a natural isomorphism $\check{H}^*_\Phi(X,F) \approx \check{H}^*_\Phi(X,F)'$. Therefore, under these conditions, we can speak of <u>the</u> Čech groups of X with coefficients in F without ambiguity.

Unfortunately, the Čech groups with coefficients in a <u>sheaf</u> do not form a δ-functor in general. If, however, we assume Φ is paracompactifying and use a suitable system of coverings (e.g., locally finite coverings or Φ-coverings), these groups will form an effaceable exact δ-functor. Consequently, under the above assumptions, the Čech groups will be isomorphic to the ordinary cohomology groups.

COVERINGS AND NERVES

Let $\mathfrak{U} = \{U_\alpha \mid \alpha \in I\}$ be a covering of X (all coverings will be indexed open coverings). We define $U_{\alpha_0,\ldots,\alpha_n} =$

$U\alpha_o \cap \cdots \cap U\alpha_n$.

Definition: The nerve $N(\mathfrak{U})$ of $\mathfrak{U}$ is the abstract simplicial complex consisting of all finite subsets $\{\alpha_o,\ldots,\alpha_n\}$ of the index set I such that $U_{\alpha_o\ldots\alpha_n} \neq \emptyset$.

If $\sigma = \alpha_o,\ldots,\alpha_n$, let $|\sigma| = U_{\alpha_o\ldots\alpha_n}$.

Now, let $\underline{S}$ be a stack over X. We define $C^n(\mathfrak{U},\underline{S})_O$ for $n \geq 0$ to be the set of functions w defined on all sequences $\alpha_o,\ldots,\alpha_n \in I$ having the property that $U_{\alpha_o\cdots\alpha_n} \neq \emptyset$ and such that $w(\alpha_o,\ldots,\alpha_n) \in \underline{S}(U_{\alpha_o\ldots\alpha_n})$. We also define $C^n(\mathfrak{U},\underline{S})_A$ to be the submodule of those w such that $w(\alpha_o \ldots \alpha_n) = 0$ if $\alpha_i = \alpha_j$ for any $i \neq j$ and such that $w(\alpha_{to},\ldots,\alpha_{tn}) = \operatorname{sgn}(t)W(\alpha_o,\ldots,\alpha_n)$ for any permutation t of $(0,\ldots,n)$. The modules $C^n(\mathfrak{U},\underline{S})_O$ and $C^n(\mathfrak{U},\underline{S})_A$ are called the modules of ordered and alternating n-cochains of $\mathfrak{U}$ with coefficients in $\underline{S}$. They are obviously K-functors in $\underline{S}$. We often omit the subscript O or A when it does not matter which module is used. As usual, we define $C^n(\mathfrak{U},\underline{S}) = 0$ for $n < 0$. However, the lemma proved below is also true if we define $C^{-1}(\mathfrak{U},\underline{S}) = \underline{S}(X)$ and $C^n(\mathfrak{U},\underline{S}) = 0$ for $n < -1$. This corresponds to augmenting $N(\mathfrak{U})$ by adding a simplex of dimension -1. I prefer to regard the coboundary $\delta: C^{-1}(\mathfrak{U},\underline{S}) \to C^o(\mathfrak{U},\underline{S})$ as an augmentation $\epsilon: \underline{S}(X) \to C^o(\mathfrak{U},\underline{S})$.

To establish its existence and naturality, however, it is preferable to regard it as part of $C(\mathfrak{U},\underline{S})$. This avoids the necessity of giving each argument twice, once for δ and once for ϵ.

There are two ways of defining supports in $C^n(\mathfrak{U},\underline{S})$.

Definition: If $w \in C^n(\mathfrak{U},\underline{S})$, then we define the support $|w|$ of w to be the closure of the set $\bigcup_{\text{all } \alpha_o,\ldots,\alpha_n} |w(\alpha_o,\ldots,\alpha_n$ Here $|W(\alpha_o,\ldots,\alpha_n)|$ means the support of $(\alpha_o,\ldots,\alpha_n) \in \underline{S}(U_{\alpha},\ldots,\alpha_n)$ defined as on p. 78 .

If Φ is any family of supports, we define

$$C^n_\Phi(\mathfrak{U},\underline{S}) = \{w \in C^n(\mathfrak{U},\underline{S}) \mid |w| \in \Phi\}.$$

We also define the support $|w|'$ of w to be the closure of the set $\bigcup U_{\alpha_o,\ldots,\alpha_n}$, the union being taken over those $(\alpha_o,\ldots,\alpha_n)$ such that $w(\alpha_o,\ldots,\alpha_n) \neq 0$. Clearly, $|w| \subset |w|'$. Therefore, if we let $C^n_\Phi(\mathfrak{U},\underline{S})' = \{w \in C^n(\mathfrak{U},\underline{S}) \mid |w|' \in \Phi\}$, we have $C_\Phi^n(\mathcal{U},\underline{S})' \subset C_\Phi^n(\mathcal{U},\underline{S})$.

We must now define the coboundary operator and maps between cochains based on different coverings. This is best done by using the theory of acyclic carriers.

Definition: Let $\mathcal{U}, \mathcal{W}$ be coverings of X. Define a carrier ψ from $N(\mathcal{W})$ to $N(\mathcal{U})$ by $\psi(\tau) = \{\sigma \in N(\mathcal{U}) \mid |\sigma| \supset |\tau|$ where τ is a simplex of $N(\mathcal{W})$.

Note that $\psi(\tau)$ is either a simplex (possibly of infinite dimension) or empty. Note also that $\psi(\tau) \neq \emptyset$ for all τ if and only if $\mathcal{W}$ is a refinement of $\mathcal{U}$. In this case, ψ is an acyclic carrier since simplexes (even infinite dimensional ones) are acyclic.

Lemma 1: Let $f: C_m(N(\mathcal{W})) \to C_n(N(\mathcal{U}))$, for some fixed m and n, be carried by ψ; i.e., $f(\sigma) \in C_n(\psi(\sigma))$ for all m-simplexes σ of $N(\mathcal{W})$.

Then f induces $f^\#: C^n(\mathcal{U},\underline{S}) \to C^m(\mathcal{W},\underline{S})$, such that $|f'(w)| \subset |w|$ and $f^\#(w)| \subset |w|$' for all w.

Furthermore, this operation gives a linear contravariant functor from the category of coverings of X and maps carried by ψ to the category of modules. A map of stacks $\underline{S} \to \underline{T}$ induces a natural transformation of functors $C(,\underline{S}) \to C(,\underline{T})$.

Note: In this lemma, C_m, C_n represent either the ordered or the alternating chain groups of $N(\mathcal{W})$, $N(\mathcal{U})$, with or without augmentation. We may even use both types at the same time, one for $\mathcal{W}$ and the other for $\mathcal{U}$. The cochain modules will, of course, be of the same type; i.e., C^n will be ordered (alternating) if C_n is ordered (alternating) and augmented if C_n is.

Proof: Let σ be an m-simplex of $N(\mathcal{W})$.

Let $f(\sigma) = \Sigma k_i t_i$ with $k_i \in K$(the ground ring) and each t_i a simplex of $N(\mathcal{U})$.

Since ψ carries f, $|t_i| \supset |\sigma|$ for all i. Therefore, we can define $f\#(w).(\sigma) = \Sigma k_i \varphi_{|\sigma|}^{|t_i|}(w(t_i))$ where $\varphi_{|\sigma|}^{|t_i|}: S(|t_i|) \to \underline{S}(|\sigma|)$ is the usual map for any stack.

It is trivial to verify that this construction has the required properties.

Remark 1: $f^\#: C_\Phi^n(\mathcal{U},\underline{S}) \to C_\Phi^m(\mathcal{W},\underline{S})$ and $f^\#: C_\Phi^n(\mathcal{U},\underline{S})' \to C_\Phi^m(\mathcal{W},\underline{S})'$

because $|f^{\#}w| \subset |w|$ and $|f^{\#}w|' \subset |w|'$

Remark 2: If $f:C(N(\mathfrak{V})) \to C(N(\mathfrak{W}))$ and $g:C(N(\mathfrak{W})) \to C(N(\mathfrak{V}))$ are both carried by ψ, then so is $fg:C(N(\mathfrak{W})) \to C(N(\mathfrak{W}))$.

Application:

a. Coboundary operator and augmentation

Let $\mathfrak{V} = \mathfrak{W}$. Then $\partial:C_n(N(\mathfrak{V})) \to C_{n-1}(N(\mathfrak{V}))$ is carried by ψ (as usual, $\partial \langle\alpha_o \dots \alpha_n\rangle = \sum_{i=o}^{n} (-1)^i \langle\alpha_o \dots \hat{\alpha}_i \dots \alpha_n\rangle$). Therefore, it induces $\delta = \partial^{\#}:C^{n-1}(\mathfrak{V},\underline{S}) \to C^n(\mathfrak{V},\underline{S})$ and $\delta^2 = 0$ since $\partial^2 = 0$. We regard $\delta:C^{-1}(\mathfrak{V},\underline{S}) \to C^o(\mathfrak{V},\underline{S})$ as an augmentation ϵ rather than as part of $C(\mathfrak{V},\underline{S})$.

Therefore, δ makes $C^{*}_{\Phi}(\mathfrak{V},\underline{S})$ a cochain complex. We define $H^{*}_{\Phi}(\mathfrak{V},\underline{S})$ to be its cohomology. The augmentation ϵ induces $\epsilon_{*}:\underline{S}_{\Phi}(X) \to H^{o}_{\Phi}(\mathfrak{V},\underline{S})$. Similarly, we define $H^{*}_{\Phi}(\mathfrak{V},S)' = H(C^{*}_{\Phi}(\mathfrak{V},\underline{S})')$. Note that $C^{-1}_{\Phi}(\mathfrak{V},\underline{S})' = 0$ unless $X \in \Phi$ so the augmentation here is of little interest in case $X \notin \Phi$.

b. Equivalence of alternating and ordered cochains

Again we let $\mathfrak{V} = \mathfrak{W}$. The theory of acyclic carriers gives us chain maps which preserve augmentation $C(N(\mathfrak{V}))_O \underset{g}{\overset{f}{\rightleftarrows}} C(N(\mathfrak{V}))_A$ and chain homotopies $f\,g \simeq id$, $gf \simeq id$, all carried by ψ. These induce corresponding cochain maps and homotopies which show $H(C^{*}_{\Phi}(\mathfrak{V},\underline{S})_A) \approx H(C^{*}_{\Phi}(\mathfrak{V},\underline{S})_O)$, this isomorphism preserving ϵ_{*}. Similarly, $H(C^{*}_{\Phi}(X,\underline{S})'_A) \approx H(C^{*}_{\Phi}(X,\underline{S})'_O)$. We can choose the map f to be the natural quotient map $C(N(\mathfrak{V}))_O \to C(N(\mathfrak{V}))_A$. A glance at the proof of the lemma then shows that $f^{\#}$ is simply the inclusion $C^{*}_{\Phi}(\mathfrak{V},\underline{S})_A \subset C^{*}_{\Phi}(\mathfrak{V},\underline{S})_O$. Thus, this inclusion is a cochain homotopy equivalence.

c. Maps from one covering to another

Let $\mathfrak{W}$ refine $\mathfrak{V}$. Then ψ is an acyclic carrier. Therefore, there are chain maps $C_{*}(N(\mathfrak{W})) \to C_{x}(N(\mathfrak{V}))$ preserving augmentation and carried by ψ. Any two are homotopic by a homotopy carried by ψ. Therefore, there are induced cochain maps $C^{*}_{\Phi}(\mathfrak{V},\underline{S}) \to C^{*}_{\Phi}(\mathfrak{W},\underline{S})$, all homotopic and preserving augmentation. Consequently, there is a unique projection $H^{*}_{\Phi}(\mathfrak{V},\underline{S}) \to H^{*}_{\Phi}(\mathfrak{W},\underline{S})$ and this preserves ϵ_{*}. Similarly, there is a unique projection $H^{*}_{\Phi}(\mathfrak{V},\underline{S})' \to H^{*}_{\Phi}(\mathfrak{W},\underline{S})'$. If $\mathfrak{M}$ is a refinement of $\mathfrak{W}$, the composition $C_{*}(N(\mathfrak{M})) \to C_{*}(N(\mathfrak{W})) \to$

$C_*(N(\mathfrak{U}))$ is again carried by ψ and so induces the projection from $H^*_\Phi(\mathfrak{U},\underline{S})$ to $H^*_\Phi(\mathfrak{V},\underline{S})$. Therefore, the projections satisfy the obvious transitivity condition and so the cohomology groups and projections form a direct system.

If $\mathfrak{V} = \{V_\beta \mid \beta \in J\}$ and $\mathfrak{U} = \{U_\alpha \mid \alpha \in I\}$, then we can find a map $r:J \to I$ such that $V_\beta \subset U_{r(\beta)}$. This map induces a simplicial map $r:N(\mathfrak{V}) \to N(\mathfrak{U})$ carried by ψ. Consequently, the projections are induced by r. This is the conventional way of defining them.

The inclusion $C^*_\Phi(\mathfrak{U},\underline{S})_A \to C^*_\Phi(\mathfrak{U},\underline{S})_O$ is natural with respect to maps induced by simplicial maps such as r. Therefore, the isomorphism $H(C^*_\Phi(\mathfrak{U},\underline{S})_A) \approx H(C^*_\Phi(\mathfrak{U},\underline{S})_O)$ is natural with respect to projections. It is trivially natural with respect to maps of stacks $\underline{S} \to \underline{T}$. The same is true of the groups $H(C^*_\Phi(\quad)')$.

We must now show that $H^*_\Phi(\mathfrak{U},\underline{S})'$ but not $H^*_\Phi(\mathfrak{U},\underline{S})$ is an exact δ-functor in $\underline{S}$ and that the projections are maps of δ-functors. Suppose we have an exact sequence $0 \to \underline{S}' \to \underline{S} \to \underline{S}'' \to 0$ of stacks. Then $0 \to C^*_\Phi(\mathfrak{U},\underline{S}')' \to C^*_\Phi(\mathfrak{U},\underline{S})' \to C^*_\Phi(\mathfrak{U},\underline{S}'')' \to 0$ is obviously a short exact sequence and so has an exact cohomology sequence. This need not be the case for $C^*_\Phi(\mathfrak{U},\underline{S})$. The inclusion $C_\Phi(\mathfrak{U},\quad)'_A \subset C_\Phi(\mathfrak{U},\quad)'_O$ gives a map of short exact sequences which then gives an isomorphism between the alternating and ordered cohomology sequences. If $\mathfrak{V}$ refines $\mathfrak{U}$, choose a chain map $f:C_*(N(\mathfrak{V})) \to C_*(N(\mathfrak{U}))$ carried by ψ and preserving augmentation. Then, $f^{\#}$ gives a map of short exact sequences which shows that the projections give a map of δ-functors.

Now, given any directed system of coverings of X, we can define $\check{H}^*_\Phi(X,S) = \varinjlim H^*_\Phi(\mathfrak{U},\underline{S})$ and $\check{H}^*_\Phi(X,\underline{S})' = \varinjlim H^*_\Phi(X,\underline{S})'$, the limit being taken over all coverings of the system. Since direct limits preserve exactness, $\check{H}^*_\Phi(X,\underline{S})'$ is obviously an exact δ-functor. Note that $\check{H}^*_\Phi(X,\underline{S})$ and $\check{H}^*_\Phi(X,\underline{S})'$ depend on the particular directed system chosen.

<u>Definition</u>: If F is a sheaf over X, let $\check{H}^*_\Phi(X,F) = \check{H}^*_\Phi(X,\Gamma(\quad,F))$ where $\Gamma(\quad,F)$ is the stack of sections of F. We also let $H^*_\Phi(\mathfrak{U},F) = H^*_\Phi(\mathfrak{U},\Gamma(\quad,F))$. Similarly, $\check{H}^*_\Phi(X,F)' = \check{H}^*_\Phi(X,\Gamma(\quad,F))'$ and $H^*_\Phi(\mathfrak{U},F)' = H^*_\Phi(\mathfrak{U},\Gamma(\quad,F))'$. Note

that this is not a δ-functor in general because Γ is not exact.

We can show, however, that $\check{H}^p_\Phi(X,\ )$ is effaceable for $p > 0$. Since every sheaf can be imbedded in a sheaf $\tilde{M}$ where M is a protosheaf, it is sufficient to prove the following lemma.

Lemma 2: Let M be a protosheaf. Let $p > 0$. Then $H^p_\Phi(\mathfrak{U},\tilde{M}) = 0$ for all coverings $\mathfrak{U}$.

Proof: For all U, $\Gamma(U,\tilde{M}) = \Gamma(U,M)$. Call this stack $\underline{S}$ for convenience. Let $\mathfrak{U} = \{U_\alpha\}$. Let $\{w_\alpha\}$ be an associated partition of unity such that

$$w_\alpha(x) = \begin{matrix} 0 \text{ for all but one } \alpha \\ 1 \text{ for some } \alpha' \text{ such that } x \in U_{\alpha'}. \end{matrix}$$

We construct w_α simply by choosing an α' for each x. Let $1_\alpha : M \to M$ be given by $1_\alpha(m) = w_\alpha(p(m)).m$. Define $D: C^n_\Phi(\mathfrak{U},\underline{S}) \to C^{n-1}_\Phi(\mathfrak{U},\underline{S})$ by $Dv.(\alpha_o,\ldots,\alpha_{n-1}) = \sum_\alpha 1_\alpha(v(\alpha,\alpha_o,\ldots,\alpha_{n-1}))$.

This is well defined. The expression $1_\alpha U(\alpha,\alpha_o,\ldots,\alpha_{n-1})$ is to be regarded as a section of M over $U_{\alpha_o\ldots\alpha_{n-1}}$. It is obviously a section over $U_{\alpha\alpha_o\ldots\alpha_{n-1}}$. But 1_α is zero over $U_{\alpha_o\ldots\alpha_{n-1}} - U_{\alpha\alpha_o\ldots\alpha_{n-1}}$. Over any point x, all but one of these sections will be zero, so the sum is defined.

Clearly, $|Du| \subset |u|$ because $|1_\alpha s| \subset |s|$ for all sections of M.

Finally, $\delta D + D\delta = id.$ in dimensions > 0. The verification of this is left to the reader. The result follows immediately from the existence of the contracting homotopy D.

Remark: Exactly the same proof shows that $H^p_\Phi(\mathfrak{U},F) = 0$ for $p > 0$ if F is fine and $\mathfrak{U}$ is locally finite.

We can also show that $\epsilon_*: \Gamma_\Phi(F) \stackrel{\approx}{\to} H^o_\Phi(X,F)$.

Lemma 3: For any sheaf F and covering $\mathfrak{U}$, $\epsilon_*: \Gamma_\Phi(F) \stackrel{\approx}{\to} H^o_\Phi(\mathfrak{U},F)$.

Proof: Let w be a 0-cochain. Then

$$\delta w.(\alpha_o,\alpha_1) = \varphi^{U_{\alpha_o}}_{U_{\alpha_o\alpha_1}}(w(\alpha_o)) - \varphi^{U_{\alpha_1}}_{U_{\alpha_o\alpha_1}}(w(\alpha_1)) = 0.$$ Therefore, the $w(\alpha) \in \Gamma(U_\alpha,F)$ are sections such that any two agree wherever they are both defined. Therefore, the sections $w(\alpha)$ amalgamate to give a section $s \in \Gamma(X,F)$. Clearly,

$\epsilon(s) = (w(\ \ ))$. Also, $\bigcup_\alpha |w(\alpha)| = |s|$ so $s \in \Gamma_\Phi(F)$. This shows ϵ_* is an epimorphism. Finally, if $t \in \Gamma_\Phi(F)$ and $\epsilon(t) = 0$, then $t | U_\alpha = 0$ for all α. Therefore, $t = 0$. Thus ϵ_* is a monomorphism.

We now show that under certain conditions, $\check{H}^*_\Phi(X,F)$ and $\check{H}^*_\Phi(X,F)'$ are naturally isomorphic.

Lemma 4: Assume Φ is such that each set of Φ has a neighborhood in Φ. Assume also that if $A, B \in \Phi$, and $A \subset \operatorname{Int} B$, Then the covering $\{X - A, \operatorname{Int} B\}$ belongs to the directed system used to define $\check{H}^*_\Phi$.

Then, $\check{H}^*_\Phi(X,F)$ and $\check{H}^*_\Phi(X,F)'$ are naturally isomorphic.

Proof: $C^*_\Phi(\mathfrak{U},\underline{S})' \subset C^*_\Phi(\mathfrak{U},\underline{S})$ for all stacks $\underline{S}$. Therefore there is a natural map $H^*_\Phi(X,F)' \to H^*_\Phi(X,F)$.

We must show that this is an isomorphism.

a. It is an epimorphism

Let $w \in C^n_\Phi(\mathfrak{U},\Gamma(\ \ ,F))$ be a cocycle, $\mathfrak{U} = \{U_\alpha \mid \alpha \in I\}$. Let $A = |w|$. Let $A \subset \operatorname{Int} B$ where $B \in \Phi$. There is a covering $\mathfrak{V} = \{V_\beta \mid \beta \in J\}$ in the system such that $\mathfrak{V}$ refines both $\mathfrak{U}$ and $\{X - A, \operatorname{Int} B\}$. Let $r{:}J \to I$ be such that $V_\beta \subset U_{r(\beta)}$. If $V_{\beta_o \ldots \beta_n} \subset X - A$, then

$$r^{\#}(w).(\beta_o \ldots \beta_n) = \varphi_{V_{\beta_o} \ldots \beta_n}^{U_{r_{\beta_o}},\ldots,r_{\beta_n}}(w(r_{\beta_o},\ldots,r_{\beta_n})).$$

But this must be zero since $V_{\beta_o \ldots \beta_n} \subset X - A$ and $A = |w|$. Consequently, $r^{\#}(w)\,(\beta_o \ldots \beta_n) \neq 0$ implies $V_{\beta_o \ldots \beta_n} \subset \operatorname{Int} B$ and so $|r^{\#}(w)|' \subset B \in \Phi$.

b. It is a monomorphism

Let $w \in C^n_\Phi(\mathfrak{U},\Gamma(\ \ ,F))'$ be such that $w = \delta v$ with $v \in C^n_\Phi(\mathfrak{U},\Gamma(\ \ ,F))$. The above argument shows that $r^{\#}(w) = r^{\#}(\delta v) = \delta\, r^{\#}(v)$ where $r^{\#}(v) \in C^n_\Phi(\mathfrak{V},\Gamma((\ \ ,F))'$. Therefore, $r^{\#}(w)$ cobounds in $C^n_\Phi(\mathfrak{V},\Gamma(\ \ ,F))$.

Let us now assume that Φ is paracompactifying and that the directed system of coverings consists either of all locally finite coverings or all Φ-coverings. Recall that a Φ-covering is a locally finite covering $\{U_\alpha \mid \alpha \in I\}$ such that there is a set $C \in \Phi$ such that $U_\alpha \subset C$ except for one exceptional value of α.

Both of these systems of coverings are obviously

directed. Furthermore, both satisfy the conditions of the previous lemma.

I claim that the defined Čech groups using a paracompactifying Φ and either of these systems of coverings form an exact δ-functor. Since all other axioms have been verified, this will show that both of these definitions give the usual cohomology groups of X with coefficients in a sheaf.

Let $0 \to F' \xrightarrow{i} F \xrightarrow{j} F'' \to 0$ be exact. Let $\overline{\Gamma}(U,F'') =$ image of $j\colon \Gamma(U,F) \to \Gamma(U,F'')$. Then $\overline{\Gamma}(\ \ ,F'')$ is a substack of $\Gamma(\ \ ,F'')$. Let $\underline{S}$ be the quotient stack. Then, in the following diagram, the horizontal and vertical sequences are exact

$$\begin{array}{ccccccccc}
 & & & & & & 0 & & \\
 & & & & & & \downarrow & & \\
0 \to & \Gamma(\ \ ,F') & \xrightarrow{i} & \Gamma(\ \ ,F) & \to & & \overline{\Gamma}(\ \ ,F'') & \to & 0 \\
 & & & & \searrow^{j} & & \downarrow & & \\
 & & & & & & \Gamma(\ \ ,F'') & & \\
 & & & & & & \downarrow & & \\
 & & & & & & \underline{S} & & \\
 & & & & & & \downarrow & & \\
 & & & & & & 0 & &
\end{array}$$

Passing to sheaves, we get

$$\begin{array}{ccccccccc}
 & & & & & & 0 & & \\
 & & & & & & \downarrow & & \\
0 \to & F' & \xrightarrow{i} & F & \to & & L\,\overline{\Gamma}(\ \ ,F'') & \to & 0 \\
 & & & & \searrow^{j} & & \downarrow & & \\
 & & & & & & F'' & & \\
 & & & & & & \downarrow & & \\
 & & & & & & S & & \\
 & & & & & & \downarrow & & \\
 & & & & & & 0 & &
\end{array}$$

Comparison with $0 \to F' \to F \to F'' \to 0$ shows that $S = 0$; i.e., that $\underline{S}$ induces the zero sheaf over X. .

Suppose we can show that $\check{H}^{*}_{\Phi}(X,\underline{S})' = 0$. Then the exact sequence for $0 \to \overline{\Gamma}(\ \ ,F'') \to \Gamma(\ \ ,F'') \to \underline{S} \to 0$ shows that $\check{H}^{*}_{\Phi}(X,\overline{\Gamma}(\ \ ,F''))' \approx \check{H}^{*}_{\Phi}(X,F'')'$. Therefore, we can substitute $\check{H}^{*}_{\Phi}(X,F'')'$ for $\check{H}^{*}_{\Phi}(X,\overline{\Gamma}(\ \ ,F''))'$ in the exact cohomology sequence of $0 \to \Gamma(\ \ ,F') \to \Gamma(\ \ ,F) \to \overline{\Gamma}(\ \ ,F'') \to 0$. This will give the required exact cohomology sequence for $0 \to F' \to F \to F'' \to 0$.

To do this, we must prove the following lemma.

Lemma 5: Let $\underline{S}$ be a stack which induces the zero sheaf on X. Let Φ be paracompactifying. Assume $H^{*}_{\Phi}(X,\underline{S})'$ is defined using either all locally finite coverings or all Φ-coverings. Then $H^{p}_{\Phi}(X,\underline{S})' = 0$ for all p.

<u>Proof</u>: Let $w \in C^p_\Phi(\mathfrak{U},\underline{S})'$ be any cochain.

Each $x \in |w|$ has a neighborhood N_x which meets only a finite number of sets U_α and so only a finite number of sets $U_{\alpha_0 \ldots \alpha_n}$. Therefore, there is a smaller neighborhood M_x such that

$$\varphi_{M_x}^{U_{\alpha_0 \ldots \alpha_n}}(w(\alpha_0 \ldots \alpha_n)) = 0 \text{ for all } \alpha_0 \ldots \alpha_n.$$

This is where we use the fact that $\underline{S}$ induces the zero sheaf. Now, if $x \notin |w|$, we define $M_x = X - |w|$. This also has the property that

$$\varphi_{M_x}^{U_{\alpha_0 \ldots \alpha_n}}(w(\alpha_0 \ldots \alpha_n)) = 0 \text{ for all } \alpha_0, \ldots, \alpha_n.$$

The $\{M_x\}$ cover X and include a set of the form X - A with $A \in \Phi$(namely $X - |w|$). Therefore it is refined by a Φ-covering $\mathfrak{V}$.

Both $\mathfrak{U}$ and $\mathfrak{V}$ belong to the directed system of coverings. Therefore, they have a common refinement $\mathfrak{W}$ in the system. The projection $r^{\#}: C^p_\Phi(\mathfrak{U},\underline{S})' \to C^p_\Phi(\mathfrak{W},\underline{S})'$ clearly annihilates w, because any $W_{\beta_0 \ldots \beta_n}$ is in some M_x and $\varphi_{M_x}^{U_{\alpha_0 \ldots \alpha_n}}(w(\alpha_0, \ldots, \alpha_n)) = 0$. Thus $\varphi_{W_{\beta_0 \ldots \beta_n}}^{U_{\alpha_0 \ldots \alpha_n}}(w(\alpha_0, \ldots, \alpha_n)) = 0$ for all $\alpha_0, \ldots, \alpha_n$ and $\beta_0, \ldots \beta_n$.

The result follows immediately from this. As an application of this theory, we prove a theorem about the Φ-dimension of X.

<u>Definition</u>: We say $\Phi\text{-dim } X \leq n$ is $H^p_\Phi(X,F) = 0$ for all $p > n$ and all sheaves F.

We say $\dim X \leq n$ if every covering of X has a refinement whose nerve has dimension $\leq n$ (as a simplicial complex).

<u>Theorem 1</u>: Let Φ be paracompactifying and such that for all $A \in \Phi$, $\dim A \leq n$. Then $\Phi\text{-dim } X \leq n$.

<u>Proof</u>: Consider the Čech cohomology groups based on Φ-coverings. It is obviously sufficient to prove the following lemma.

<u>Lemma 6</u>: Let Φ be paracompactifying and such that for all $A \in \Phi$, $\dim A \leq n$. Then every Φ-covering of X is refined by a Φ-covering whose nerve has dimension $\leq n$.

<u>Proof</u>: Let $\mathfrak{U} = \{U_\alpha \mid \alpha \in I\}$ be a Φ-covering of X. Then there is a set $C \in \Phi$ and $\alpha_0 \in I$ such that $U_\alpha \subset C$ for all $\alpha \neq \alpha_0$.

Now, $\{U_\alpha \cap C \mid \alpha \in I\}$ covers C. Since dim $C \leq n$, it has a refinement $\{V_\beta \mid \beta \in J\}$ with a nerve of dimension $\leq n$. Choose $r: J \to I$ such that $V_\beta \subset U_{r(\beta)}$. Define a new covering $\mathscr{W} = \{W_\alpha \mid \alpha \in I\}$ of C by $W_\alpha = \bigcup V_\beta$, the union being taken over all β such that $r(\beta) = \alpha$. Clearly, the nerve of $\mathscr{W}$ has dimension $\leq n$. Also, $W_\alpha \subset U_\alpha$. Therefore, $\mathscr{W}$ is locally finite.

Finally, we define a covering $\mathscr{G} = \{G_\alpha \mid \alpha \in I\}$ of X by $G_\alpha = W_\alpha$ if $\alpha \neq \alpha_o$, $G_{\alpha_o} = W_{\alpha_o} \cup (X - C)$. Then $\mathscr{G}$ is a locally finite open Φ-covering of X, it refines $\mathscr{U}$, and its nerve has dimension $\leq n$.

To conclude this account of the Čech theory, we will consider cup products.

Let $\underline{S}$ and $\underline{T}$ be stacks over X. Let $\Phi = \Phi_1 \cap \Phi_2$. Define maps $\mu: C^p_{\Phi_1}(\mathscr{U}, \underline{S}) \otimes C^q_{\Phi_2}(\mathscr{U}, \underline{T}) \to C^{p+q}_\Phi(\mathscr{U}, \underline{S} \otimes \underline{T})$ by $\mu(w \otimes v)$.
$(\alpha_o \ldots \alpha_{p+q}) = \varphi^{U_{\alpha_o \ldots \alpha_p}}_{U_{\alpha_o \ldots \alpha_{p+q}}} w(\alpha_o \ldots \alpha_p) \otimes \varphi^{U_{\alpha_o \ldots \alpha_{p+q}}}_{U_{\alpha_o \ldots \alpha_{p+q}}}$
$v(\alpha_p \ldots \alpha_{p+q})$.

These maps are easily seen to be well defined and to be cochain maps. They commute with projections of the form $r^\#$ where r is the usual map of the index set of $\mathscr{W}$ to the index set $\mathscr{U}$. Therefore, they induce the cup product maps,

$$\check{H}^p_{\Phi_1}(X, \underline{S}) \otimes \check{H}^q_{\Phi_2}(X, \underline{T}) \to \check{H}^{p+q}_\Phi(X, \underline{S} \otimes \underline{T})$$

and $$\check{H}^p_{\Phi_1}(X, \underline{S})' \otimes \check{H}^q_{\Phi_2}(X, \underline{T})' \to \check{H}^{p+q}_\Phi(X, \underline{S} \otimes \underline{T})' .$$

These maps are obviously consistent with the natural maps $C_\Phi(\ \ )' \to C_\Phi(\ \ )$ and $H_\Phi(\ \ )' \to H_\Phi(\ \ \ )$. It is also obvious that

$$\begin{array}{ccc} \underline{S}_{\Phi_1}(X) \otimes \underline{T}_{\Phi_2}(X) & \to & (\underline{S} \otimes \underline{T})_\Phi(X) \\ \downarrow \epsilon_* \otimes \epsilon_* & & \downarrow \epsilon_* \\ \check{H}^o_{\Phi_1}(X, \underline{S}) \otimes \check{H}^o_{\Phi_2}(X, \underline{T}) & \to & \check{H}^o_\Phi(X, \underline{S} \otimes \underline{T}) \end{array}$$

is commutative.

If $0 \to \underline{S}' \to \underline{S} \to \underline{S}'' \to 0$, $0 \to \underline{R}' \to \underline{R} \to \underline{R}'' \to 0$ are exact and

$$\begin{array}{ccccccc} \underline{S}' \otimes \underline{T} & \to & \underline{S} \otimes \underline{T} & \to & \underline{S}'' \otimes \underline{T} & \to & 0 \\ \downarrow & & \downarrow & & \downarrow & & \\ 0 \to \underline{R}' & \to & \underline{R} & \to & \underline{R}'' & \to & 0 \end{array}$$

is commutative, it is trivial to verify that

$$\begin{array}{ccc} \check{H}^p_{\Phi_1}(X,\underline{S}'')' \otimes \check{H}^q_{\Phi_2}(X,\underline{T})' & \xrightarrow{\delta \otimes 1} & \check{H}^{p+1}_{\Phi_1}(X,\underline{S}')' \otimes \check{H}^q_{\Phi_2}(X,\underline{T})' \\ \downarrow & & \downarrow \\ \check{H}^{p+q}_{\Phi}(X,\underline{R}'')' & \xrightarrow{\delta} & \check{H}^{p+q+1}_{\Phi}(X,\underline{R}')' \end{array}$$

is commutative. If we tensor on the left with $\underline{T}$, there is a similar diagram but with the usual change of sign; i.e., the top map should be $w \otimes \delta$, not $1 \otimes \delta$.

Now, if Φ is paracompactifying and if Čech groups are defined using locally finite or Φ-coverings, the above properties show that the cup products satisfy the usual axioms and so agree with the usual cup products. Details are left to the reader.

IX. THE SPECTRAL SEQUENCES

PRELIMINARIES

The spectral sequences will be defined in terms of double complexes. We will first remind the reader of some properties of these complexes. Full details may be found in CE, chapter IV, section 4, and chapter XV, section 6.

Definition: A double complex is a bigraded $A = \sum_{i,j} A^{i,j}$ together with an endomorphism $\delta : A \to A$ such that

(1) $\delta^2 = 0$

and (2) $\delta = \delta_1 + \delta_2$ where $\deg \delta_1 = (1,0)$ and $\deg \delta_2 = (0,1)$ (By $\deg f = (p,q)$, I mean $f: A^{i,j} \to B^{i+p,j+q}$ for all i,j. Here f is any map $f: A \to B$ of bigraded objects.)

To every double complex, there is associated a cochain complex with $A^n = \sum_{i+j=n} A^{i,j}$ and with $\delta : A^n \to A^{n+1}$ given by the δ of the double complex. We call n the "total degree". We define $H^n(A)$ to be the nth cohomology module of this complex.

There are also two other ways to make A into a cochain complex. We may take A with the coboundary operator δ_1 or with the coboundary operator δ_2.

Define: $H_I^{p,q}(A)$ = cohomology of A with respect to δ_1

$H_{II}^{p,q}(A)$ = " " " " " δ_2

Furthermore, $\delta_1\delta_2 = -\delta_2\delta_1$. Therefore δ_2 induces a coboundary operator δ_{2*} on $H_I(A)$. Consequently, we may define $H_{II}^{p,q}\, H_I(A)$ = cohomology of $H_I(A)$ with respect to δ_{2*}. In exactly the same way, we define $H_I^{p,q} H_{II}(A)$.

If A is a double complex, we can filter A in two ways as follows:

$$F_I^{\,p}(A) = \sum_{\substack{i \geq p \\ j}} A^{i,j} \qquad \text{"the first filtration"}$$

$$F_{II}^{\,q}(A) = \sum_{\substack{j \geq q \\ i}} A^{i,j} \qquad \text{"the second filtration"}$$

Each of these filtrations makes A a filtered differential graded object, graded by total degree, and with δ as differential operator. Consequently, each of those filtrations yields a spectral sequence. We refer to these sequences as the first and second spectral sequence, respectively. In the first spectral sequence

$$E_0^{i,j} = A^{i,j},\ E_1^{i,j} = H_{II}^{i,j}(A),\ E_2^{i,j} = H_I^{i,j}\ H_{II}(A)$$

and $E_\infty^{i,j} = \mathcal{G}^{i,j}\ H^{i+j}(A)$, the graded object associated with a filtration of H(A) induced by F_I on A.

For the second spectral sequence, similar results hold but with the subscripts I and II interchanged.

A map of double complexes is a map which preserves bigrading (i.e., has degree(0,0)) and commutes with δ. Clearly, the various objects and spectral sequences are functors on the category of double complexes and maps.

If $f,g:A \to B$ are maps of double complexes, we say f and g are homotopic ($f \simeq g$) when there is a map $s:A \to B$ such that

(1) $s = s_1 + s_2$ where $\deg s_1 = (-1,0)$, $\deg s_2 = (0,-1)$ and (2) $f - g = \delta s + s\delta$.

It is not hard to show that if $f \simeq g$, the induced maps of spectral sequences are the same except on the terms E_0 and E_1 of each spectral sequence.

Finally, we recall some sufficient conditions for the convergence of the spectral sequences.

1. If for some fixed M, we have $A^{i,j} = 0$ for $i < -M$, then the second filtration is regular; i.e., for each fixed n, $H^n(F_{II}^p A) = 0$ if p is large enough. Therefore the second spectral sequence will converge; i.e., $E_\infty^{i,j} = \varinjlim_k E_k^{i,j}$.
2. If for some fixed N we have $A^{i,j} = 0$ for $j < -N$, then the first filtration is regular and the first spectral sequence will converge.
3. If both the preceding conditions are satisfied, both spectral sequences converge. In fact, in this case both satisfy the following condition:

For fixed i,j, and large k, $E_k^{i,j} = E_\infty^{i,j}$.

4. Suppose that for some fixed M and N, we have $A^{i,j} = 0$ for $i < -M$ or $i > N$, j being unrestricted. Then again both filtrations will be regular, both spectral sequences will converge, and, in fact, there will be a k_o such that $E_k^{i,j} = E_\infty^{i,j}$ for all $k \geq k_o$ and all i,j.

A similar result holds if $A^{i,j} = 0$ for $j < -M$ or $j > N$ and i is unrestricted.

To construct the spectral sequences we start out with a covariant, left-exact K-functor T: $\mathcal{A} \to \mathcal{B}$ and assume that there are natural T-resolutions and resolvent functors over T. We refer the reader to chapter III, section 4 for the definitions and properties of these. In chapter IV we have shown that if $\mathcal{A}$ is the category of sheaves, $\mathcal{B}$ that of modules, and $T = \Gamma_\Phi$ there are natural T-resolutions (in fact, the canonical flabby resolution is one such) and resolvent functors over T. Certainly these are the categories we have in mind throughout this chapter. The double complexes and spectral sequences will always be in the category $\mathcal{B}$.

In order to discuss the convergence of the spectral sequences, we first give some criteria for there to be finite dimensional natural T-resolutions.

Definition: A natural resolution N is of dimension $\leq$ n if $N^i(A) = 0$ for $i > n$, all $A \in \mathcal{A}$.

A resolvent functor $\mathcal{F}$ over T is of dimension $\leq$ n if $\mathcal{F}^i(A) = 0$, $i > n$, $A \in \mathcal{A}$.

Definition: A category $\mathcal{A}$ is of T-dimension $\leq$ n if $R^iT.(A) = 0$, $i > n$, $A \in \mathcal{A}$. When $\mathcal{A}$ is the category of sheaves over X, $T = \Gamma_\Phi$, we express this definition by saying that Φ-dim $X \leq n$.

Proposition 1: Let T be left-exact. Then the following statements are equivalent.

1. T-dim $\mathcal{A} \leq n$.
2. $R^{n+1}T.(A) = 0$ for all $A \in \mathcal{A}$.
3. There is a natural T-resolution of dim $\leq$ n.
4. There is a resolvent functor over T of dim $\leq$ n.

Proof: (1) $\Rightarrow$ (2) trivially.

(3) $\Rightarrow$ (4) by proposition 8, chapter III.

(4) $\Rightarrow$ (1) by corollary to proposition 9, chapter III.

To complete the proof show (1) $\Rightarrow$ (3) and (2) $\Rightarrow$ (4).

To show (1) $\Rightarrow$ (3). Let N' be a natural T-resolution. Define a new one N by $N^i = N'^i$ if $i < n$, $N^i = 0$, $i > n$, and $N^n(A) = Z^n N'(A)$.

The maps δ and ϵ for N are induced by those for N'.

N is a natural T-resolution. For $0 \to A \to N^0(A) \to N^1(A) \to \ldots$ is exact for all A. If $0 \to A' \to A \to A'' \to 0$ is exact, so is $0 \to N(A') \to N(A) \to N(A'') \to 0$.

For since $H^{n+1}(N'(A')) = 0$, lemma 5, p. 50 shows that $0 \to Z^n N'(A') \to Z^n N'(A) \to Z^n N'(A'') \to 0$ is exact. Therefore N^n is an exact functor. The other N^i obviously are.

We verify this is a T-resolution; i.e., that $R^i T . N^j(A) = 0$, $i > 0$, all j, A. This is obviously true for $j \neq n$. For $j = n$, and $p \geq 1$ lemma 6, p. 53 shows that $R^p T .\ N^n(A) \approx R^{p+n} T.(A) = 0$.

To show (2) $\Rightarrow$ (4). We take any resolvent functor $\mathcal{F}'$ over T. Define $\mathcal{F}$ by $\mathcal{F}^i = \mathcal{F}'^i$, if $i < n$, $\mathcal{F}^i = 0$, if $i > n$, and $\mathcal{F}^n = Z^n \circ \mathcal{F}$ (i.e., $\mathcal{F}^n(A) = \ker \delta \mid \mathcal{F}'^n(A)$). The maps δ and ϵ for $\mathcal{F}$ are induced by those of $\mathcal{F}'$.

Properties (ii) and (iii) of resolvent functors (cf. p. 54) are obviously satisfied by $\mathcal{F}$. To prove (i), let $0 \to A' \to A \to A'' \to 0$ be exact. Then $0 \to \mathcal{F}'(A') \to \mathcal{F}'(A) \to \mathcal{F}'(A'') \to 0$ is exact. Apply lemma 5, p. 50 using the fact that $H^{n+1}(\mathcal{F}'(A')) = R^{n+1}T.(A') = 0$. This shows that $\mathcal{F}^n = Z^n \circ \mathcal{F}$ is exact. The remaining $\mathcal{F}^i$ are obviously exact. This construction of $\mathcal{F}$ from $\mathcal{F}'$ will be used again later.

Note that lemma 6, p. 53 and the existence of injective imbeddings show directly that (2) $\Rightarrow$ (1).

THE SPECTRAL SEQUENCES

I will define the spectral sequences using a resolvent functor and show that they are independent of the choice of this functor. At the end of this section, I will prove that these spectral sequences agree with those defined by CE chapter XVII. This fact will not be used in the applications however.

When I say two spectral sequences are isomorphic, I will mean that they are isomorphic from the term E_2 on. It does not matter if the E_o and E_1 terms are isomorphic or not.

Let M be a cochain complex of elements of $\mathcal{Q}$, i.e., a sequence $\ldots \xrightarrow{\delta} M^q \xrightarrow{\delta} M^{q+1} \xrightarrow{\delta} \ldots$ such that $\delta\delta = 0$.

I will not assume in general that $M^q = 0$ for all q less than some q_o. If this condition is satisfied, I will say that M is bounded below.

Let $\mathfrak{F}$ be a resolvent functor for T. If we know that T-dim $\mathcal{Q} < \infty$, we choose $\mathfrak{F}$ to be finite dimensional. Define a double complex $\mathfrak{F}(M)$ as follows:

$\mathfrak{F}(M)^{p,q} = \mathfrak{F}^p(M^q)$ with $\delta = \delta_1 + \delta_2$ where

(a) $\delta_1 : \mathfrak{F}^p(M^q) \to \mathfrak{F}^{p+1}(M^q)$ is the map δ given by the definition of a resolvent functor, and

(b) $\delta_2 : \mathfrak{F}^p(M^q) \to \mathfrak{F}^p(M^{q+1})$ is $(-1)^p \mathfrak{F}^p(\delta)$, with the δ here being the δ of M.

We also define a map $\epsilon : T(M) \to \mathfrak{F}(M)$ to be the map given by the augmentation of $\mathfrak{F}$; i.e., $\epsilon : T(M^q) \to \mathfrak{F}^o(M^q)$ for all q. It is trivial to verify that ϵ is a cochain map.

The two spectral sequences associated with M are now defined to be the two spectral sequences of the double complex $\mathfrak{F}(M)$. They are obviously K-functors in M. The second sequence always converges (since $\mathfrak{F}^p = 0$ for $p < 0$). The first will converge if M is bounded below or if we know that T-dim $\mathcal{Q} < \infty$ (recall that $\mathfrak{F}$ is assumed finite dimensional whenever T-dim $\mathcal{Q}$ is known to be finite).

Obviously, $H_I(\mathfrak{F}(M)) = RT.(M)$ and $H_{II}(\mathfrak{F}(M)) = \mathfrak{F}(H(M))$ where H(M) is the cohomology of the complex M.

Therefore, in the first spectral sequence, we have:

$E_o^{i,j} = \mathfrak{F}^i(M^j)$, $E_1^{i,j} = \mathfrak{F}^i(H^j(M))$, and $E_2^{i,j} = R^iT.(H^j(M))$

In the second spectral sequence,

$E_o^{i,j} = \mathfrak{F}^i(M^j)$, $E_1^{i,j} = R^iT(M^j)$, and $E_2^{i,j} = H^j(R^iT.(M))$, this last being the cohomology of $R^iT.(M)$ considered as a cochain complex with coboundary operator δ_* induced by δ acting on M.

The E_∞ terms of the two sequences will be the graded modules associated with the two filtrations of $H(\mathfrak{F}(M))$. We call $H(\mathfrak{F}(M))$ the hypercohomology module of M. In order to apply the spectral sequences, it is necessary to compute this module.

Proposition 2:

Suppose $H^q R^p T.(M) = 0$ for all $p \neq 0$ and all q. (Note that this is just $E_2^{p,q}$ of the second sequence). Then $\epsilon_* : H(T(M)) \to H(\mathfrak{F}(M))$ is an isomorphism.

Proof:

Bigrade T(M) by $T(M)^{o,q} = T(M^q)$ and $T(M)^{p,q} = 0$ if $p \neq 0$. Then T(M) with its usual δ (induced by that of M) becomes a double complex and ϵ becomes a map of doubles complexes. Since the second spectral sequences of T(M) and $\mathfrak{F}(M)$ both converge, it is sufficient to show that ϵ induces an isomorphism of E_2 terms of these sequences.

Now, $E_1^{p,q}(\mathfrak{F}(M)) = R^p T.(M)$ while $E_1^{p,q}(T(M))$ is T(M) if p = 0 and zero if $p \neq 0$. The map $E_1^{o,q}(T(M)) \to R^o T.(M)$ is just the usual augmentation $T(M) \to R^o T.(M)$. Since this is an isomorphism, so is $E_2^{o,q}(T(M)) \to E_2^{o,q}(\mathfrak{F}(M))$ (since deg. $d_1 = (0,1)$). But, if $p \neq 0$, $E_2^{p,q} = 0$ in both sequences.

We can use this proposition to show that the spectral sequences are independent of the choice of $\mathfrak{F}$, at least if M is bounded below or T-dim $\mathfrak{a}$ is known to be finite.

Suppose $\mathfrak{F}$ is a resolvent functor for T; and C is a natural T-resolution. Consider $\mathfrak{F} \circ C$. This is a functor which assigns to every F the double complex $\mathfrak{F}(C(F))$ with ϵ: $T(C(F)) \to \mathfrak{F}(C(F))$ as above. We can think of $\mathfrak{F}(C(F))$ as a cochain complex graded by the total degree in the usual way. Define $\epsilon : T(F) \to \mathfrak{F}(C(F))$ to be the composition $T(F) \xrightarrow{T(\epsilon)} T(C(F)) \xrightarrow{\epsilon} \mathfrak{F}(C(F))$.

Proposition 3: $\mathfrak{F} \circ C$, with the ϵ just defined, is a resolvent functor for T.

Proof:

Clearly $\delta\epsilon = 0$ and $\delta\delta = 0$. Also $\mathfrak{F} \circ C$ is an exact functor, being the composition of two exact functors. Now, the previous proposition applies with M = C(F) since $C^i(F)$ is T-acyclic for all i. Therefore, $\epsilon_* : H(T(C(F))) \to H(\mathfrak{F} \circ C(F))$ is an isomorphism for all F. But, $T \circ C$ is a resolvent functor. Therefore $H^i(\mathfrak{F} \circ C(F)) = 0$ for $i > 0$ if

F is injective. Finally $\epsilon : T(F) \to R^0T(F)$ is an isomorphism, being the composition of the isomorphisms $T(F) \overset{\approx}{\to} H^0(T(C(F)))$ $\underset{\approx}{\overset{\epsilon_*}{\to}} H^0(\mathfrak{F} \circ C(F))$.

Corollary:

If $\mathfrak{F}$, $\mathfrak{F}'$ are any two resolvent functors for T, there are resolvent functors $\mathfrak{F}_0$, $\mathfrak{F}_1$, and $\mathfrak{F}_2$ for T and maps $\mathfrak{F} \to \mathfrak{F}_0 \leftarrow \mathfrak{F}_1 \to \mathfrak{F}_2 \leftarrow \mathfrak{F}'$. If $\mathfrak{F}$ and $\mathfrak{F}'$ all have dimension $\leq$ n, we can choose $\mathfrak{F}_0$, $\mathfrak{F}_1$, $\mathfrak{F}_2$ to have dimension $\leq$ n.

Proof: Let C be a natural T-resolution.

Consider

$$\mathfrak{F}(F) \xrightarrow{\mathfrak{F}(\epsilon)} \mathfrak{F}(C(F)) \overset{\epsilon_*}{\leftarrow} T(C(F)) \overset{\epsilon'_*}{\to} \mathfrak{F}'(C(F)) \xleftarrow{\mathfrak{F}'(\epsilon)} \mathfrak{F}'(F).$$

This has the required properties except for the dimension condition.

Now, if dim $\mathfrak{F} \leq$ n, then T-dim $\mathcal{A} \leq$ n. Consequently, if $\mathcal{G}$ is any resolvent functor, we get a new resolvent functor by taking $\mathcal{G}^i$ in dimensions $i < n$, $Z^n(\mathcal{G})$ in dimension n, and 0 in dimensions $> n$ (with δ and ϵ induced by that of $\mathcal{G}$). Apply this process to all five functors in $\mathfrak{F} \to \mathfrak{F}_0 \leftarrow \mathfrak{F}_1 \to \mathfrak{F}_2 \leftarrow \mathfrak{F}'$. It leaves $\mathfrak{F}$ and $\mathfrak{F}'$ unchanged and replaces $\mathfrak{F}_0$, $\mathfrak{F}_1$, and $\mathfrak{F}_2$ by resolvent functors of dimension $\leq$ n.

Corollary:

Let $M \in \mathcal{C}\mathcal{A}$. Let $\mathfrak{F}$ and $\mathfrak{F}'$ be resolvent functors. Assume either that M is bounded below or that $\mathfrak{F}$ and $\mathfrak{F}'$ are finite dimensional. Then the spectral sequences of M based on $\mathfrak{F}$ and $\mathfrak{F}'$ are naturally isomorphic.

Proof: By the preceding corollary, it is sufficient to prove this for the case in which there is a map $\mathfrak{F} \to \mathfrak{F}'$. But, such a map will obviously induce isomorphisms of the E_2 terms because any maps of resolvent functors induces a natural isomorphism $H \circ \mathfrak{F} \overset{\approx}{\to} H \circ \mathfrak{F}'$.

Remark: If $\mathfrak{F}$, $\mathfrak{F}'$, and M are unrestricted, the same proof shows that the spectral sequences are isomorphic except possibly for the E_∞ terms of the first sequence. The only difficulty encountered is, of course, the non-convergence of the first sequences.

Proposition 4: Let M and M' be cochain complexes of sheaves.

Let f: $M \to M'$ be a cochain map. Let $\mathfrak{F}$ be any resolvent functor. Assume that either T-dim $\mathfrak{a} < \infty$ or that both M and M' are bounded below.

Then, if $f_* : H(M) \to H(M')$ is an isomorphism so is $f_\# : H(\mathfrak{F}(M)) \to H(\mathfrak{F}(M'))$.

Note that $\mathfrak{F}$ need not be finite dimensional even if M and M' are not bounded below.

Proof: By the remark above, $H(\mathfrak{F}(M))$ and $H(\mathfrak{F}(M'))$ are independent of the choice of $\mathfrak{F}$, being limits of the second spectral sequence. Consequently, we can replace $\mathfrak{F}$ by a finite dimensional resolvent functor in case T-dim $\mathfrak{a} < \infty$.

Now, consider the first spectral sequences. These will converge. Since f_* obviously gives an isomorphism at E_2, the result follows immediately.

Corollary: Let M, M', and f satisfy the conditions of the proposition. Assume in addition that $HR^pT.(M) = 0$ and $HR^pT.(M') = 0$ for $p \neq 0$.

Then $f_\# : H(T(M)) \to H(T(M'))$ is an isomorphism.

Proof: This is an immediate consequence of the proposition the naturality of ϵ_*, and the fact that ϵ_* is an isomorphism for M and M'.

A natural map: Let M be a cochain complex. Assume T-dim $\mathfrak{a} < \infty$ or M bounded below. Assume also that $H^q(M) = 0$ for $q > k$ (some given k).

Under these conditions, I will define a natural map $\alpha : H^{p+k}(T(M)) \to R^pT.H^k(M)$. This is done as follows:

Let $M_1{}^q$ be M^q if $q < k$; $Z^k(M)$ if $q = k$; and zero if $q > k$. Then M_1 is a subsomplex of M. Let $\beta : M_1 \to M$ be the inclusion. Let $\gamma : M_1 \to H^k(M)$ be zero on all $M_1{}^q$ with $q \neq k$. If $q = k$, then $M_1{}^q = Z^k(M)$ and $\gamma \mid M_1{}^q$ is defined to be the natural epimorphism. Note that we can regard $H^k(M)$ as a cochain complex zero in every dimension but k. If we do this, γ becomes a cochain map.

Now, consider the maps $H^{p+k}(T(M)) \overset{\epsilon_*}{\to} H^{p+k}(\mathfrak{F}(M)) \overset{\beta_\#}{\leftarrow} H^{p+k}(\mathfrak{F}(M_1)) \overset{\gamma_\#}{\to} H^{p+k}(\mathfrak{F}(H^k(M)) = R^pT.H^k(M)$. Since $\beta_* : H(M_1) \to H(M)$ is an isomorphism, $\beta_\#$ will also be an isomorphism because of the conditions imposed on M. Therefore we can define α to be $\gamma_\# \circ \beta_\#^{-1} \circ \epsilon_*$.

To show α is independent of the choice of $\mathfrak{F}$, it is sufficient (by corollary p. 118) to consider the case where α is defined using resolvent functors $\mathfrak{F}$ and $\mathfrak{F}'$ for which there is a map $f: \mathfrak{F} \to \mathfrak{F}'$. This f induces a map of the above sequence of maps ϵ_*, $\beta_\#$ and $\gamma_\#$ for $\mathfrak{F}$ to the same sequence for $\mathfrak{F}'$. It gives the natural isomorphisms of the modules at the ends. Consequently, α is independent of the choice of $\mathfrak{F}$.

Proposition 5: Assume

(1) T-dim $\mathcal{a} < \infty$ or M bounded below,

(2) $HR^pT(M) = 0$ for $p \neq 0$, and

(3) $H^q(M) = 0$ for $q \neq k$.

Then $\alpha : H^{p+k}(T(M)) \to R^pT.H^k(M)$ is an isomorphism for all p.

Proof:

By (2), ϵ_* is an isomorphism.

By (1), $\beta_\#$ is an isomorphism.

Finally, (3) implies that γ, considered as a cochain map, gives an isomorphism $H(M) \to H(H^k(M)) = H^k(M)$. This, together with (1), implies that $\gamma_\#$ is an isomorphism.

We can also show that α preserves δ whenever δ is defined. Suppose $0 \to M' \xrightarrow{i} M \xrightarrow{j} M'' \to 0$ is an exact sequence of cochain maps. Define $\overline{T}(M'')$ to be the image of $T(j) : T(M) \to T(M'')$. Therefore, $0 \to T(M') \to T(M) \to \overline{T}(M'') \to 0$ is exact. Note that, in spite of the notation, $\overline{T}(M'')$ depends on j as well as on M''. If α is defined for M'', we define α': $H^{p+k}(\overline{T}(M'')) \to R^pT.H^k(M'')$ to be the composition of α with the map $H(\overline{T}(M'')) \to H(T(M''))$ induced by inclusion.

The sequence $0 \to T(M') \to T(M) \to \overline{T}(M'') \to 0$ gives an exact sequence of cohomology groups. If, in addition, we assume $0 \to H^k(M') \to H^k(M) \to H^k(M'') \to 0$ is exact, we can apply the δ-functor R^*T to get another exact sequence.

Proposition 6: Let $0 \to M' \xrightarrow{i} M \xrightarrow{j} M'' \to 0$ be an exact sequence of cochain maps.

Assume either that T-dim $\mathcal{a} < \infty$ or that all three complexes M', M and M'' are bounded below. Suppose also that $H^q(M') = H^q(M) = H^q(M'') = 0$ for $q > k$ and that $0 \to H^k(M') \to H^k(M) \to H^k(M'') \to 0$ is exact.

Then the following diagram commutes:

$$\begin{array}{ccccccccc}
\cdots\to H^{p+k}(T(M')) & \to & H^{p+k}(T(M)) & \to & H^{p+k}(\overline{T}(M'')) & \xrightarrow{\delta} & H^{p+k+1}(T(M)) \to \cdots \\
\downarrow & & \downarrow & & \downarrow & & \downarrow \\
\cdots\to R^pT.H^k(M') & \to & R^pT.H^k(M) & \to & R^pT.H^k(M'') & \to & R^{p+1}T.H^k(M') \to \cdots
\end{array}$$

Proof:

The first two squares commute by the naturality of α and the definition of α'. To show the third commutes, we first notice that $0 \to M_1' \to M_1 \to M_1'' \to 0$ is exact. This follows immediately from lemma III. 5, p. 50 and the fact that $H^{k+1}(M') = 0$.

We now consider the following commutative diagrams with exact rows:

1.
$$\begin{array}{ccccccccc}
0 & \to & T(M') & \to & T(M) & \to & \overline{T}(M'') & \to & 0 \\
 & & \downarrow\epsilon & & \downarrow\epsilon & & \downarrow\epsilon & & \\
0 & \to & \mathfrak{I}(M') & \to & \mathfrak{I}(M) & \to & \mathfrak{I}(M'') & \to & 0
\end{array}$$

2.
$$\begin{array}{ccccccccc}
0 & \to & M' & \to & M & \to & M'' & \to & 0 \\
 & & \uparrow\beta & & \uparrow\beta & & \uparrow\beta & & \\
0 & \to & M_1' & \to & M_1 & \to & M_1'' & \to & 0
\end{array}$$

3.
$$\begin{array}{ccccccccc}
0 & \to & M_1' & \to & M_1 & \to & M_1'' & \to & 0 \\
 & & \downarrow & & \downarrow & & \downarrow & & \\
0 & \to & H^k(M_1') & \to & H^k(M_1) & \to & H^k(M_1'') & \to & 0
\end{array}$$

We leave (1) as it is, but apply $\mathfrak{I}$ to (2) and (3). We then get maps of cohomology sequences from the resulting diagrams. The "squares" involving δ fit together as follows

$$\begin{array}{ccccccc}
H(\overline{T}(M'')) & \xrightarrow{\epsilon_*} & H(\mathfrak{I}(M'')) & \xleftarrow{\beta_\#} & H(\mathfrak{I}(M_1'')) & \xrightarrow{\gamma_\#} & H(\mathfrak{I}(H^k(M''))) \\
\downarrow\delta & & \downarrow\delta & & \downarrow\delta & & \downarrow\delta \\
H(T(M')) & \xrightarrow{\epsilon_*} & H(\mathfrak{I}(M')) & \xleftarrow{\beta_\#} & H(\mathfrak{I}(M_1')) & \xrightarrow{\gamma_\#} & H(\mathfrak{I}(H^k(M')))
\end{array}$$

This diagram shows that $\alpha\delta = \delta\alpha'$.

Corollary: If α is an isomorphism for M' and M, then α' is also an isomorphism (by the 5-lemma).

COMPARISON WITH THE CARTAN-EILENBERG THEORY

The results of this section will not be used anywhere else in these notes. Consequently the reader interested only in the applications may skip this section.

In chapter XVII of CE, spectral sequences are defined for a cochain complex M and a functor such as T. These spectral sequences are defined in terms of an injective resolution of M. The only property of this resolution we use here is that it is a T-resolution in the sense of the following definition.

Definition: A T-resolution of a cochain complex of elements of $\mathcal{A}$ (where T is as usual a covariant left-exact K-functor $\mathcal{A} \to \mathcal{B}$) consists of a couble complex C of elements of $\mathcal{A}$ and a map $\epsilon : M \to C$ such that

(1) ϵ is a cochain map;

(2) $\epsilon : M^q \to C^{o,q}$;

(3) $C^{p,q} = 0$ for $p < 0$;

(4) For each q;

$$0 \to M^q \xrightarrow{\epsilon} C^{o,q} \xrightarrow{\delta} C^{1,q} \xrightarrow{\delta} \ldots \quad \text{and}$$
$$0 \to H^q(M) \xrightarrow{\epsilon_*} H_{II}{}^{o,q}(C) \xrightarrow{\delta_{1*}} H_{II}{}^{1,q}(C) \xrightarrow{\delta_{1*}} \ldots$$

are exact; and

(5) All $C^{p,q}$, $Z_{II}{}^{p,q}(C)$, $B_{II}{}^{p,q}(C)$, and $H_{II}{}^{p,q}(C)$ are T-acyclic (i.e., annihilated by R^iT for $i > 0$).

Let $M \xrightarrow{\epsilon} C$ and $M' \xrightarrow{\epsilon'} C'$ be T-resolutions, and $f: M \to M'$ a cochain map. A map of T-resolutions covering f is a map of double complexes $g: C \to C'$ such that

$$\begin{array}{ccc} C & \xrightarrow{g} & C' \\ \epsilon \uparrow & & \uparrow \epsilon' \\ M & \xrightarrow{f} & M' \end{array}$$

commutes.

We say the T-resolution C is regular if either

(a) For some q_o, $C^{p,q} = 0$ for all $q < q_o$

or (b) For some p_o, $C^{p,q} = 0$ for all $p > p_o$.

If C is such a resolution, T(C) is a double complex. Its spectral sequences are the sequences considered by Cartan-Eilenberg (for the case in which C is an injective resolution). The second sequence always converges, and the first converges if C is regular.

Let $\mathfrak{F}$ be a resolvent functor for T. Then $\mathfrak{F}(C)$ is a triple complex where:

$\mathfrak{F}(C)^{p,q,r} = \mathfrak{F}^p(C^{q,r})$ and $\delta = \delta_1' + \delta_2' + \delta_3'$ where

(a) δ_1' is the map $\mathfrak{F}^p(C^{q,r}) \to \mathfrak{F}^{p+1}(C^{q,r})$ given by the definition of a resolvent functor;

(b) $\delta_2' : \mathfrak{F}^p(C^{q,r}) \rightarrow \mathfrak{F}^p(C^{q+1,r})$ is $(-1)^p \mathfrak{F}^p(\delta_1)$; and
(c) $\delta_3' : \mathfrak{F}^p(C^{q,r}) \rightarrow \mathfrak{F}^p(C^{q,r+1})$ is $(-1)^p \mathfrak{F}^p(\delta_2)$.

We convert $\mathfrak{F}(C)$ into a double complex by defining

$$\mathfrak{F}(C)^{p,q} = \sum_{p'+p''=p} \mathfrak{F}(C)^{p',p'',q}$$

Define $\epsilon_1 : T(C) \rightarrow \mathfrak{F}(C)$ to be the map given by $\epsilon : T(C^{p,q}) \rightarrow \mathfrak{F}^o(C^{p,q})$.

Define $\epsilon_2 : \mathfrak{F}(M) \rightarrow \mathfrak{F}(C)$ to be the map $\mathfrak{F}(\epsilon)$ where $\epsilon : M \rightarrow C$.

It is trivial to verify that ϵ_1 and ϵ_2 are maps of double complexes.

The required isomorphisms between the spectral sequences based on $\mathfrak{F}$ and those of Cartan-Eilenberg follow immediately from the next proposition. More precisely, they follow from it if C is regular and either $\mathfrak{F}$ is finite dimensional or M is bounded below. If these conditions are not satisfied, the sequences will be isomorphic with the possible exception of the E_∞ terms of the first sequences.

Proposition 7: ϵ_1 and ϵ_2 induce isomorphisms of the E_2 terms of the first spectral sequences of $T(C)$, $\mathfrak{F}(C)$, and $\mathfrak{F}(M)$. They induce isomorphisms of the E_1 terms of the second spectral sequences.

Proof:

We first consider the second spectral sequences. The operator δ_2 is not used in computing E_1. Consequently, we can ignore δ_2 and keep the second index q fixed.

Now, if $N^p = C^{p,q}$ (with q fixed) and $\delta : N^q \rightarrow N^{q+1}$ is δ_1 for C, then N satisfies the conditions of proposition 2. Therefore $\epsilon_* : H^p(T(N)) \rightarrow H^p(\mathfrak{F}(N))$ is an isomorphism, but this is just the map $\epsilon_1 : E_1^{p,q} \rightarrow E_1^{p,q}$ for the second spectral sequences.

The definition of T-resolution (in particular, property (4) implies that $\epsilon : M \rightarrow H_I(C)$ is an isomorphism. Therefore, we consider M^q (fixed q) as a cochain complex zero except in dimension 0, the map $\epsilon : M^q \rightarrow N$ (where N is as before) induces an isomorphism $H(M^q) \rightarrow H(N)$. Therefore, by proposition 4, $\epsilon_\# : H^p(\mathfrak{F}(M^q)) \rightarrow H^p(\mathfrak{F}(N))$ is an isomorphism. But, this is $\epsilon_2 : E_1^{p,q} \rightarrow E_1^{p,q}$ for the second spectral sequences.

We now consider the first spectral sequences. By property (4) of T-resolutions, $\epsilon_*: H(M) \to H_{II}(C)$ induces an isomorphism $\epsilon_*: H(M) \to H_I H_{II}(C)$. Now, the E_1 terms for $\mathfrak{F}(M)$ and $\mathfrak{F}(C)$ are $\mathfrak{F}(H(M))$ and $\mathfrak{F}(H_{II}(C))$. Therefore we can apply proposition 4 to show that $E_2 = H(\mathfrak{F}(H(M)) \xrightarrow{\epsilon_2} H(\mathfrak{F}(H_{II}(C)) = E_2$ is an isomorphism.

Finally, $H_{II}(C)$ satisfies the conditions of proposition 2. Therefore $\epsilon: T(H_{II}(C)) \to \mathfrak{F}(H_{II}C)$ induces an isomorphism $H(T(H_{II}(C))) \to H(\mathfrak{F}H_{II}C))$. To show $\epsilon_1: E_2 \to E_2$ is an isomorphism, it is now sufficient to show that $E_1(T(C)) = T(H_{II}(C))$. In doing this, we can ignore the index p and the operator δ_1. Therefore it is sufficient to prove the following lemma.

Lemma 1:

Let M be a cochain complex of sheaves. Assume all M^q, $Z^q(M)$, $B^q(M)$, and $H^q(M)$ are T-acyclic. Then $H^q(T(M)) \approx T(H^q(M))$ naturally, and in such a way that the following diagram commutes:

$$\begin{array}{ccc} H(T(M)) & \approx & T(H(M)) \\ \downarrow \epsilon_* & & \downarrow \epsilon \\ H(\mathfrak{F}(M) & \approx & \mathfrak{F}(H(M)) \end{array}$$

Proof: The T-acyclicity implies that if we apply T to any of the exact sequences

$$0 \to B^q \to Z^q \to H^q \to 0$$

and $0 \to Z^q \to C^q \to H^q \to 0$,

the result will be an exact sequence. We now use exactly the same proof which is used to show that $H(T(M)) = T(H(M))$ for any exact functor T. At the same time, we repeat this proof for the exact functor $\mathfrak{F}$. Since the steps used for T and $\mathfrak{F}$ are the same, it is clear that

$$\begin{array}{ccc} H(T(M)) & \to & T(H(M)) \\ \downarrow \epsilon_* & & \downarrow \epsilon \\ H(\mathfrak{F}(M)) & \to & \mathfrak{F}(H(M)) \end{array}$$

commutes.

X. THE SPECTRAL SEQUENCE OF A MAP

Lemma 1:

Let M and M' be cochain complexes of sheaves. Let $f,g:M \to M'$ be cochain homotopic cochain maps. Then f and g induce the same maps of the spectral sequences of M and M'.

Proof: Let $f - g = \delta s + s\delta$. Define

$$s' = (-1)^p \mathfrak{F}^p(s): \mathfrak{F}^p(M) \to \mathfrak{F}^p(M').$$

Then s' is a homotopy between the maps of double complexes $\mathfrak{F}(f)$ and $\mathfrak{F}(g)$.

We will apply this lemma as follows:

Let S be a linear functor from sheaves over a space E to sheaves over a space B. Let G be a sheaf over E. Let I be an injective resolution of G. Then S(I) is a cochain complex of sheaves and is bounded below. The functor Γ_Φ with Φ a family of supports in B, induces spectral sequences of S(I). If $f: G \to G'$ is a map of sheaves, take injective resolutions I and I' of G and G' and a map $g: I \to I'$ covering f. Then g is unique up to homotopy. Therefore, given f, there is a unique induced map of spectral sequences. The same arguments used in defining derived functors now show that the spectral sequences are independent (up to natural isomorphism) of the choice of the resolution I and that the sequences are linear functors in G.

The term E_2 of the first sequence will be $E_2^{i,j} = H^i_\Phi(B, H^j(S(I)))$. But, by definition, $H^j(S(I)) = R^jS.(G)$. Therefore, in the first spectral sequence, $E_2^{i,j} = H^i_\Phi(B, R^jS.(G))$.

We can also find $H^i(\Gamma_\Phi(S(I)))$. By definition this is just $R^i(\Gamma_\Phi \circ S).(G)$. This is mapped by ϵ_* into the hypercohomology module of S(I). Therefore, by using the proposition which gives a sufficient condition for ϵ_* to be an isomorphism, we get the following lemma.

Lemma 2:

Suppose that S(F) is Φ-acyclic for every injective sheaf F over E. Then there is a natural spectral sequence, defined for all sheaves G over E, with

$$E_2^{i,j}(G) = H_\Phi^i(B, R^j S.(G))$$

and

$$E_\infty^{i,j}(G) = \mathcal{G}^{i,j}[R^{i+j}(\Gamma_\Phi \circ S).G.]$$

Proof: Since S(I) is Φ-acyclic, $H_\Phi^p(B, S(I)) = 0$ for $p \neq 0$. Therefore ϵ_* is an isomorphism.

To get the spectral sequence of a map $f: E \to B$, we apply this lemma with an appropriate choice of S. Throughout this section, f will denote a fixed map $f: E \to B$.

Suppose that for each open U in B we are given a family of supports Ψ(U) in $f^{-1}(U)$ satisfying the following condition:

If $A \in \Psi(U)$ and $V \subset U$, then $A \cap f^{-1}(V) \in \Psi(V)$. Then the modules $\Gamma_{\Psi(U)}(f^{-1}(U), G)$ with the obvious restriction maps form a stack over B.

Definition: $f_\Psi(G)$ is the sheaf over B defined by this stack. Obviously, f_Ψ is a left-exact, linear, covariant functor.

Definition: The spectral sequence of f (with the given Φ and Ψ) is the first spectral sequence derived from the functor $S = f_\Psi$ from sheaves on E to sheaves on B, where the T of chapter IX is Γ_Φ, for a family Φ in B.

We must now calculate E_2 and E_∞. In particular, we must calculate $\Gamma_\Phi \circ f_\Psi$ and give conditions under which $f_\Psi(F)$ is Φ-acyclic for all injective F.

Definition: For each open $U \subset B$, let Ψ'(U) be the collection of sets $A \subset f^{-1}(U)$ such that

(1) A is (relatively) closed in $f^{-1}(U)$ and

(2) Every $x \in U$ has an open neighborhood N_x such that $A \cap f^{-1}(N_x) \in \Psi(N_x)$.

Remark: If we repeat this construction twice, we get nothing new; i.e., $\Psi''(U) = \Psi'(U)$ for all U. This is obvious from the definition of Ψ'(U).

Lemma 3:

There is a natural isomorphism of stacks $\Gamma(U, f_\Psi(G)) \approx$

$\Gamma_{\Psi'(U)}(f^{-1}(U),G)$. If $s \in \Gamma(U,f_\Psi(G))$ corresponds to $t \in \Gamma_{\Psi'(U)}(f^{-1}(U),G)$, then the support $|s|$ of s is the closure of $f(|t|)$. Here $|s|$ and $|t|$ are the supports of s and t considered as sections of sheaves not as elements of stacks.

<u>Proof</u>: For each U, $\Psi(U) \subset \Psi'(U)$. Therefore there is a map of stacks $\Gamma_{\Psi(U)}(f^{-1}(U),G) \to \Gamma_{\Psi'(U)}(f^{-1}(U),G)$.

I claim that this map induces an isomorphism of sheaves. Since the map of stacks is a monomorphism, it induces a monomorphism of sheaves. Now, let $t \in \Gamma_{\Psi'(U)}(f^{-1}(U),G)$ and let $x \in U$. Then x has a neighborhood N_x such that $|t| \cap f^{-1}(N_x) \in \Psi(N_x)$. This shows that the image of t in $\Gamma_{\Psi'(N_x)}(f^{-1}(N_x),G)$ is actually in $\Gamma_{\Psi(N_x)}(f^{-1}(N_x),G)$. This result clearly implies that the map of sheaves is an epimorphism.

To prove the lemma, it is now sufficient to show that the stack $\Gamma_{\Psi'(U)}(f^{-1}(U),G)$ has the collation property and no locally zero elements. Let U be open in B and $\{U_\alpha\}$ a covering of U. If $t \in \Gamma_{\Psi'(U)}(f^{-1}(U),G)$ has image zero in each $\Gamma_{\Psi'(U_\alpha)}(f^{-1}(U_\alpha),G)$, then $t \mid f^{-1}(U_\alpha) = 0$ for all α. Since t is a section of sheaf G, $t = 0$. Therefore the stack has no locally zero elements. Now, let U and $\{U_\alpha\}$ be as above and suppose elements $t_\alpha \in \Gamma_{\Psi'(U_\alpha)}(f^{-1}(U_\alpha),G)$ are given such that t_α and t_β have the same image in $\Gamma_{\Psi'(U_\alpha \cap U_\beta)}(f^{-1}(U_\alpha \cap U_\beta),G)$ for all α and β. Then $t_\alpha \mid f^{-1}(U_\alpha) \cap f^{-1}(U_\beta) = t_\beta \mid f^{-1}(U_\alpha) \cap f^{-1}(U_\beta)$ for all α and β. Therefore, the t_α's fit together to give a section $t \in \Gamma(f^{-1}(U),G)$. Obviously, $|t| \cap f^{-1}(U_\alpha) = |t_\alpha| \in \Psi'(U)$.

Therefore $|t| \in \Psi''(U) = \Psi'(U)$.

Finally, let $s \in \Gamma(U,f_\Psi(G))$ correspond to $t \in \Gamma_{\Psi'(U)}(f^{-1}(U),G)$. Let x be any point of U. Then $x \notin |s|$ if and only if x has a neighborhood N_x such that $s \mid N_x = 0$. This is equivalent to saying $t \mid f^{-1}(N_x) = 0$ which in turn is equivalent to saying $|t| \cap f^{-1}(N_x) = \emptyset$ or $f(|t|) \cap N_x = \emptyset$. Therefore, $|s|$ = closure in U of $|t|$.

<u>Corollary</u>: Let Ψ'' be the collection of sets A such that $A \in \Psi(B)$ and $\overline{f(A)} \in \Phi$. Then $\Gamma_\Phi \circ f_\Psi = \Gamma_{\Psi''}$.

Corollary: $R^i(\Gamma_\Phi f_\Psi) = R^i\Gamma_{\Psi''} = H^i_{\Psi''}(E, \ )$.

Corollary:

$f_{\Psi'} = f_\Psi$.

We must now find conditions under which $f_\Psi(I)$ is Φ-acyclic for all injective I.

Lemma 4:

1. Suppose that for every U, $\Psi(U)$ consists of all (relatively) closed sets of $f^{-1}(U)$. Then $f_\Psi(I)$ is injective whenever I is injective.
2. Suppose Φ is PF (no restriction on $\Psi(U)$). Then $f_\Psi(F)$ is Φ-soft whenever F is flabby.

Proof:

1. Let I be injective. For each $x \in E$, choose an injective module $J_x \supset I_x$. Then the J_x form a protosheaf $\bar{J}$ and $I \subset \tilde{\bar{J}}$. Since I is injective, it is a direct summand. But, f_Ψ is a linear functor; therefore, $f_\Psi(I)$ is a direct summand of $f_\Psi(\tilde{\bar{J}})$, and so it is sufficient to show $f_\Psi(\tilde{\bar{J}})$ is injective.

   Now, if V is open in E, $\Gamma(V,\tilde{\bar{J}}) = \Gamma(V,\bar{J})$. Therefore $\Gamma_{\Psi(U)}(f^{-1}(U),\tilde{\bar{J}}) = \Gamma(f^{-1}(U),\tilde{\bar{J}}) = \Gamma(f^{-1}(U),\bar{J}) = \Gamma(U,\bar{N})$ where $\bar{N}$ is the protosheaf over B defined by $N_b = \prod_{y \in f^{-1}(b)} J_y$. Therefore, $f_\Psi(\tilde{\bar{J}}) = \tilde{\bar{N}}$, but this is injective because each N_b is injective (being a direct product of injective modules).
2. Let F be flabby, Since $f_{\Psi'} = f_\Psi$, it is sufficient to show that $f_{\Psi'}(F)$ is Φ-soft.

   Let $A \in \Phi$ and $g \in \Gamma(A,f_\Psi(F))$. Since Φ is PF, we can find a neighborhood A' of A in Φ and g extends to an open neighborhood U of A in A' because A' is paracompact. Let $g' \in \Gamma(U,f_\Psi(F))$ be this extended section. Choose a set $B \in \Phi$ such that $B \subset U$ and $A \subset \operatorname{Int} B$. This can be done since A' is paracompact and therefore normal. We write V = Int B.

   Now, let $h \in \Gamma_{\Psi(U)}(f^{-1}(U),F)$ be the element corresponding to g' under the canonical isomorphism given by the previous lemma. Then

$|h| \cap f^{-1}(B)$ is closed in E since $|h|$ is closed in $f^{-1}(U)$, $f^{-1}(B) \subset f^{-1}(U)$, and $f^{-1}(B)$ is closed in E.

Let h' be the section of F which is h over $f^{-1}(V)$ and zero outside $|h| \cap f^{-1}(B)$. Since h' is obviously continuous and F is flabby, we can extend h' to a section h" of F over E. Now, $|h''| \subset |h| \cap f^{-1}(B)$. Since $|h| \in \Psi'(B)$, it is easy to see that $|h''| \in \Psi''(B) = \Psi'(B)$. Therefore $h \in \Gamma_{\Psi(B)}(E,F)$. The corresponding element in $\Gamma(B,f_\Psi(F))$ is clearly an extension of g to B.

Corollary: Assume either

(1) For all U, $\Psi(U)$ = all (relatively) closed sets of $f^{-1}(U)$ <u>or</u>

(2) Φ is PF.

Then, in the spectral sequence of f with respect to Φ and Ψ, we have $E_\infty^{i,j} = \mathcal{G}^{i,j}\, H^{i+j}_{\Psi''}(E,G)$ where Ψ'' was defined in the first corollary of the previous lemma.

Proof: The present lemma implies that ϵ_* is an isomorphism. Therefore $E_\infty^{i,j} = \mathcal{G}^{i,j}\, R^{i+j}(\Gamma_\Phi \circ f_\Psi)$ but, $R^{i+j}\, \Gamma_\Phi \circ f_\Psi = H^i_{\Psi''}(E,\ \ )$ by the second corollary of the previous lemma.

We now must determine the E_2 term. We have already seen that $E_2^{i,j} = H^i_\Phi(B, R^j f_\Psi.(G))$. Now, for each j, the cohomology groups $H^j_{\Psi(U)}(f^{-1}(U), G \mid f^{-1}(U)$ form a stack with maps induced by the inclusions $f^{-1}(V) \to f^{-1}(U)$ for $V \subset U$.

Lemma 5:

For all j, $R^j f_\Psi.(G)$ is naturally isomorphic to the sheaf defined by the stack $H^j_{\Psi(U)}(f^{-1}(U), G \mid f^{-1}(U))$.

Proof: By proposition VII, 3, p. 95 , the $H^j_{\Psi(U)}(f^{-1}(U), (\ \ ) \mid f^{-1}(U)$ for a cohomological δ-functor over $\Gamma_{\Psi(U)}(f^{-1}(U), (\ \ )$ since $f^{-1}(U)$ is open for all open U. Since the functor L from stacks to sheaves is exact, the sheaves $L\,[H^i_{\Psi(U)}(f^{-1}(U), (\ \ ) \mid f^{-1}(U)]$ form a cohomological δ-functor over $L(\Gamma_{\Psi(U)}(f^{-1}(U),\ \ )) = f_\Psi$. The result follows from the uniqueness of cohomological δ-functors.

Under certain conditions, we can find a simple expression for the stalk $f_\Psi(G)_x$.

Lemma 6:

Assume that E is locally compact and that E and B are Hausdorff. For each U, let $\Psi(U)$ be the collection of all intersections of $f^{-1}(U)$ with compact sets of E.

Then, $R^j f_\Psi(G)_x = H^j_{\Psi_x}(f^{-1}(x), G \mid f^{-1}(x))$ for all x, where Ψ_x is the set of compact subsets of $f^{-1}(x)$.

Proof: This follows immediately from the previous lemma and the continuity property proved in chapter VII, p. 98, Ex. 2 We have only to verify that $\bigcap_{U \ni x} f^{-1}(U) = f^{-1}(x)$. But, $\bigcap_{U \ni x} f^{-1}(U) \subset \bigcap_{U \ni x} f^{-1}(\overline{U}) = f^{-1}(\bigcap_{U \ni x} \overline{U})$. Since B is Hausdorff, $\bigcap_{U \ni x} \overline{U} = x$.

We will now examine the relation between the spectral sequence and the map f^* of cohomology induced by $f: E \to B$. In order to define f^* it is necessary to assume that $A \in \Phi$ implies $f^{-1}(A) \in \Psi''$. We write $f^{-1}\Phi \subset \Psi''$ for short. We also assume that every set of Φ has a neighborhood in Φ. This implies that the union W of all sets of Φ is open and that every point of W has a neighborhood in Φ.

Let G be a sheaf over B, and $f^{-1}(G)$ the induced sheaf over E. Suppose Φ and Ψ'' satisfy the conditions just mentioned. Then the natural map $\Gamma(U,G) \to \Gamma(f^{-1}(U), f^{-1}(G))$ induces a map $\Gamma_{\Phi(U)}(U,G) \to \Gamma_{\Psi'(U)}(f^{-1}(U), f^{-1}(G))$ where $\Phi(U)$ is collection of intersections $A \cap U$ of U with sets A of Φ.

Lemma 7:

Suppose every set of Φ has a neighborhood in Φ. Let W be the union of all sets in Φ. Let $\Phi(U) = \{A \cap U \mid A \in \Phi\}$. Then the stack $\Gamma_{\Phi(U)}(U,G)$ defines the sheaf G_W and the inclusion $\Gamma_{\Phi(U)}(U,G) \to \Gamma(U,G)$ induces the usual inclusion $G_W \to G$

Proof: Let F be the sheaf defined by $\Gamma_{\Phi(U)}(U,G)$. The monomorphisms $\Gamma_{\Phi(U)}(U,G) \to \Gamma(U,G)$ induce a monomorphism $F \to G$. We only have to show that $F_x \to G_x$ is an epimorphism for $x \in W$ and that $F_x = 0$ for $x \notin W$.

If $x \in W$, then x has a neighborhood $N_x \in \Phi$. If $U \subset N_x$, then $\Phi(U)$ is the collection of all (relatively) closed subsets of U. Therefore $\Gamma_{\Phi(U)}(U,G) = \Gamma(U,G)$. Consequently, $F_x = G_x$ for $x \in W$.

Now, suppose $x \notin W$. Let $s \in \Gamma_{\Phi(U)}(U,G)$ with $x \in U$. Then $|s| \in \Phi(U)$ and so, $|s| = U \cap A$ with $A \in \Phi$.

But $A \subset W$. Therefore $x \notin A$. Since A is closed, there is an open set V such that $x \in V \subset U$ and $V \cap A = \emptyset$. Therefore, $s \mid V = 0$. Since s was arbitrary, $F_x = 0$.

Corollary: Suppose every set of Φ has a neighborhood in Φ and that $f^{-1}(\Phi) \subset \Psi''$. Let W be the union of all sets in Φ. Then there is a natural map $i\colon G_W \to f_\Psi(f^{-1}(G))$ for all sheaves G over B.

Proof: This map is induced by the map of stacks $\Gamma_{\Phi(U)}(U,G) \to \Gamma_{\Psi(U)}(f^{-1}(U), f^{-1}(G))$.

Proposition 1: Suppose every set of Φ has a neighborhood in Φ and that $f^{-1}(\Phi) \subset \Psi''$. Let W be the union of all sets in Φ. Let G be any sheaf over B. The set of cohomology groups $H^i_\Phi(B,G)$ can be regarded as forming a trivial spectral sequence $E'^{i,j}_k$ with $E'^{i,j}_k = 0$ for $j \neq 0$ and $E'^{i,j}_k = H^i_\Phi(B,G)$ for all $j = 0$ and all i and k.

Under the above conditions, there is a natural map of this spectral sequence into the spectral sequence of f with coefficients in $f^{-1}(G)$.

The map of limit groups given by this map is the following composition

$$H^i_\Phi(B,G) \xrightarrow{f^*} H^i_{\Psi''}(E,f^{-1}(G)) \xrightarrow{\epsilon^*} H(\mathfrak{F} f_\Psi(I)).$$

The map of E_2 terms is given by the composition

$$H^i_\Phi(B,G) \overset{\approx}{\leftarrow} H^i_\Phi(B,G_W) \to H^i_\Phi(B,f_\Psi f^{-1}(G)) = E_2^{i,0}(f^{-1}(G)).$$

Note that the map on the left, induced by $G_W \subset G$, is an isomorphism since $\Phi = \Phi_W$. The equality on the right follows from the fact that f_Ψ is left exact and so $R^0 f_\Psi = f_\Psi$. The map in the middle is induced by the natural map $i\colon G_W \to f_\Psi f^{-1}(G)$.

Proof: Let J be an injective resolution of G. Let I be an injective resolution of $f^{-1}(G)$. Then, $f^{-1}(J)$ is an acyclic complex over $f^{-1}(G)$ (f^{-1} being an exact functor) so there is a cochain map $f^{-1}(J) \to I$ which preserves augmentation. This map is unique up to homotopy. By applying f_Ψ and using the map i, the augmentations, and the inclusions $G_W \subset G$ etc, we get the following sequence of maps:

$$G \leftarrow G_W \xrightarrow{\epsilon} J_W \xrightarrow{i} f_\Psi\; f^{-1}(J) \to f_\Psi(I)$$

If we regard G and G_W as cochain complexes, zero except

in dimension 0, this diagram is composed of cochain maps of cochain complexes over B. Each of these complexes has a spectral sequence. Those for G and G_W are trivial and both are isomorphic to $H_\Phi(B,G)$. That of $f_\Psi(I)$ is the spectral sequence of f. The maps induced by the cochain maps in the diagram give the required map of spectral sequences. It is easy to see, in the usual way, that this map is independent of the choice of I and J and is natural.

If we take the cohomology of the cochain complexes in the above sequence, we get, in dimension zero,

$$G \leftarrow G_W \overset{\equiv}{\to} G_W \to f_\Psi\, f^{-1}(G) \overset{\equiv}{\to} f_\Psi\, f^{-1}(G).$$

To get the last two terms, we use the fact that f_Ψ is left-exact. To find the map of E_2 terms, we apply $H^i_\Phi(B, \ \ )$ to this sequence. The result is clearly as stated in the proposition.

Note that J_W is an acyclic complex over G_W. Therefore, $H(G_W) \to H(J_W)$ is an isomorphism. This shows that the first spectral sequences of G_W and J_W are isomorphic. Furthermore, J_W is flabby since J, being injective, is flabby and W is open. Therefore $H^p_\Phi(B,J_W) = 0$ for $p \neq 0$ and so, ϵ_*: $H\,\Gamma_\Phi(J_W) \to H\,\Gamma_\Phi(\mathfrak{F}(J_W))$ is an isomorphism.

Consider the diagram

$$\begin{array}{ccccccccc} \mathfrak{F}(G) \leftarrow \mathfrak{F}(G_W) \to & \mathfrak{F}(J_W) & \to & \mathfrak{F}(f_\Psi f^{-1}(J)) & \to & \mathfrak{F}f_\Psi(I) & \\ & \uparrow\epsilon & & \uparrow\epsilon & & \uparrow & \\ & \Gamma_\Phi(J_W) & \to & \Gamma_\Phi f_\Psi f^{-1}(J)) & \to & \Gamma_\Phi f_\Psi(I) & = \Gamma_{\Psi''}(I). \end{array}$$

This and the above remarks show that the map of limit groups can be factored as follows:

$$H(\mathfrak{F}(G)) \to H(\Gamma_{\Psi''}(I)) \overset{\epsilon_*}{\to} H(\mathfrak{F}f_\Psi(I))$$

Now, the map on the left is a map of augmented δ-functors. To see this, suppose we have a short exact sequence of G's. Then, since f^{-1} is exact, we also have a short exact sequence of $f^{-1}(G)$'s. We can find short exact sequences of injective resolutions over each of these. Such sequences split in every dimension, so the result of applying any linear functor is again a short exact sequence. Therefore, the maps in the above diagram give us maps of short exact sequences and so, maps of their cohomology sequences. This shows that δ is preserved. To see that ϵ is preserved, we take the above diagram and replace I by $f^{-1}(G)$, J by G and $\mathfrak{F}$ by Γ_Φ. This gives us a new diagram which is mapped into the old one by means of the augmentations $G \to J$, $f^{-1}(G) \to I$ and $\Gamma_\Phi \to \mathfrak{F}$. In this new diagram, the map corresponding to $H(\mathfrak{F}(G)) \to H(\Gamma_{\Psi''}(I))$ is just the usual map $\Gamma_\Phi(G) \to \Gamma_{\Psi''}(f^{-1}(G))$.

Since the map $H(\mathfrak{F}(G)) \to H(\Gamma_{\Psi''}(I))$ preserves ϵ and δ, it must be the unique map $f^*: H_\Phi(B,G) \to H_{\Psi''}(E, f^{-1}(G))$. Therefore, the map of limit groups is as stated in the proposition.

As an application of the spectral sequence of a map, we now prove the Vietoris mapping theorem.

<u>Theorem</u>: Let $f: E \to B$ where E and B are Hausdorff and E is locally compact. For each $x \in B$ let Ψ_x be the collection of compact sets in $f^{-1}(x)$. Let Φ be any PF family of supports in B. Let $\Psi(H)$ be as in lemma 6.

Let G be a constant sheaf over E. Assume that for all $x \in B$, we have $H^i_{\Psi_x}(f^{-1}(x),G) \overset{\epsilon}{\approx} G$ for $i = 0$ and $= 0$ for $o < i < k$, where k is some fixed integer independent of x. Then, for $i < k$ there is an isomorphism $H^i_{\Psi''}(E,G) \approx H^i_\Phi(B,G)$. Also, there is a monomorphism $H^k_\Phi(B,G) \to H^k_{\Psi''}(E,G)$.

If we assume that $f^{-1}(\Phi) \subset \Psi''$, then these maps of cohomology are just the maps f^* induced by f.

<u>Proof</u>: In the spectral sequence of f, we have $E_2^{i,j} = 0$ for $0 < j < k$ and $E_2^{i,o} = H^i_\Phi(B,G)$. This follows immediately from lemma 6. The conclusion now follows immediately from standard spectral sequence arguments.

Suppose now that $f^{-1}(\Phi) \subset \Psi''$ so that f^* is defined. Consider the map of spectral sequences given by proposition 1. Clearly, $G_W \to f_\Psi f^{-1}(G)$ is an isomorphism in stalks over W since this is just $\epsilon: G \to H^o_{\Psi_x}(f^{-1}(x),G) = R^o f_{\Psi}.(G) = f_\Psi(G)$. Therefore, the map is an isomorphism on $E_2^{i,o}$ for all i. But, in both sequences, $E_2^{i,j} = 0$ if $0 < j < k$. Therefore, the map is also an isomorphism for these values of k. Consequently, the map of limit groups is an isomorphism in dimensions $< k$ and a monomorphism in dimension k. But, this map is just $\epsilon_* f^*$ and ϵ_* is an isomorphism in all dimensions since Φ is PF.

XI. THE DUALITY THEOREMS

<u>SINGULAR HOMOLOGY</u>

Let X be any space. Let S be the sheaf of singular chains over X with coefficients in the ground ring K.

<u>Definition</u>: If G is any sheaf over X, we define $H_i^{\Phi}(X,G) = H_i(\Gamma_{\Phi}(S \otimes G))$.

<u>Remark</u>: If G is a constant sheaf, we have seen that $\Gamma_{\Phi}(S \otimes G)$ is naturally isomorphic to the usual complex of singular chains with coefficients in G and supports in Φ. Therefore $H_i^{\Phi}(X,G)$ agrees with the classical singular homology groups if G is constant and Φ is the collection of compact subsets of X. (X being assumed Hausdorf here.)

Suppose U is an open subset of X. Let A = X - U. Let G be any sheaf over X. Then $0 \to G_U \to G \to G_A \to 0$ is exact and so is $0 \to S \otimes G_U \to S \otimes G \to S \otimes G_A \to 0$ (in fact, $S \otimes G_U = (S \otimes G)_U$ and $S \otimes G_A = (S \otimes G)_A$). Let $\overline{\Gamma}_{\Phi}(S \otimes G_A)$ be the image of $\Gamma_{\Phi}(S \otimes G)$ in $\Gamma_{\Phi}(S \otimes G_A)$.

<u>Definition</u>: $H_i^{\Phi}(X,U;G) = H_i(\overline{\Gamma}_{\Phi}(S \otimes G_A))$

<u>Remark</u>: The short exact sequence

$$0 \to \Gamma_{\Phi}(S \otimes G_U) \to \Gamma_{\Phi}(S \otimes G) \to \overline{\Gamma}_{\Phi}(S \otimes G_A) \to 0$$

gives an exact homology sequence,

$$\ldots \to H_i(\Gamma_{\Phi}(S \otimes G_U)) \to H_i^{\Phi}(X,G) \to H_i^{\Phi}(X,U;G) \xrightarrow{\partial} \ldots .$$

We want to try and identify $H_i(\Gamma_{\Phi}(S \otimes G_U))$ with $H_i^{\Phi_U}(U,G \mid U)$ and show that if G is constant, the above sequence is just the ordinary homology sequence of the pair (X,U). This can be done using the spectral sequences provided Φ - dim U $< \infty$ and Φ is PF.

<u>Definition</u>: $S^i = S_{-i}$ for all i.

This gives a new grading for S making it a cochain complex. We can consider the first spectral sequence of this complex, or, more generally, of $S \otimes G$ for any coefficient sheaf G. Since S is homotopically fine, so is $S \otimes G$. The required endomorphisms corresponding to a covering U_{α} are just $\{1_\alpha \otimes 1\}$ where $1_\alpha : S \to S$ are the endomorphisms for S. If Φ is PF, this implies that $H\, H^p_\Phi(X,S) = 0$ for $p \neq 0$ and so, $\epsilon_*: H(\Gamma_\Phi(S \otimes G)) \approx H(\mathcal{F}(S \otimes G))$ for all G. Therefore, the limit group of the spectral sequence is $H^\Phi_i(X,G)$. The E_2 term is $E_2^{i,j} = H^i_\Phi(X, H^j(S \otimes G))$. Note that if X is Hausdorff, $S_x = \underline{S}(X,X-x)$, the usual singular chain complex of the pair (X,X-x). Therefore, $H^j(S \otimes G)_x = H_{-j}(X,X-x;\, G_x)$ where the right hand side is just the classical singular homology group.

Note that S, considered as a cochain complex, is not bounded below so we must assume Φ - dim $X < \infty$ to make the spectral sequence converge.

<u>Lemma 1</u>:

Let $U \subset X$ be open. Assume Φ is PF and that Φ_U - dim U $< \infty$. Then $H_i(\Gamma_\Phi(S \otimes G_U))$ and $H_i^{\Phi_U}(U,G \mid U)$ are naturally isomorphic for all i and all sheaves G over X. If G is constant, then the exact sequence

$$\ldots \to H_i(\Gamma_\Phi(S \otimes G_U)) \to H^\Phi_i(X,G) \to H^\Phi_i(X,U;G) \to \ldots$$

is naturally isomorphic to the classical homology sequence of the pair (X,U)(using singular homology with supports in Φ).

<u>Proof</u>: Let S^+ be the sheaf of singular chains over U. Then S^+ is defined by the stack $\underline{S}(U,U-V)$ where V is any open set of U. But $S \mid U$ is defined by the stack $\underline{S}(X,X-V)$. Therefore the inclusion $\underline{S}(U,U-V) \to \underline{S}(X,X-V)$ induces a map of sheaves $S^+ \to S \mid U$.

Now, let W be the union of all sets in Φ. Then W is open, and every point of W has a neighborhood in Φ. I claim that if $x \in W \cap U$ then $H(S^+ \otimes G)_x \to H(S \otimes G)_x$ is an isomorphism. To see this, let N_x be any open neighborhood of x in U such that $\overline{N}_x \in \Phi$. Since $\overline{N}_x$ is paracompact, it is regular, so we can find an open neighborhood M_x such that $\overline{M}_x \subset N_x$. We can now **excise** $U - N_x$ from $(U,U-M_x)$ and $X - N_x$ from $(X,X-M_x)$. The result in both cases is $(N_x,\, N_x - M_x)$. Therefore, the excision axiom shows that $H(U,U - M_x;G_x) \to H(X,X - M_x;G_x)$ is an isomorphism. The result now follows from the fact that taking direct limits commutes with taking homology so that $H(S^+ \otimes G)_x = H(S^+_x \otimes G_x) = \varinjlim H(\underline{S}(U,U-M_x;G_x)) \approx$, etc.

To prove the lemma, we regrade S as explained above and consider the spectral sequences (over U) of S^+ and $S \mid U$. For any sheaf F, over U, $H_{\Phi_U}(U,F)$ depends only on $F_{W \cap U}$ because $W \cap U$ contains the union of all sets of Φ_U. Therefore, the map $S^+ \otimes G \mid U \to S \mid U \otimes G \mid U$ induces an isomorphism of E_2 terms of the first spectral sequence and hence an isomorphism

$$H(\Gamma_{\Phi_U}(S^+ \otimes G \mid U) \to H(\Gamma_{\Phi_U}((S \otimes G) \mid U)).$$

But, $\Gamma_{\Phi_U}((S \otimes G) \mid U) = \Gamma_{\Phi}((S \otimes G)_U) = \Gamma_{\Phi}(S \otimes G_U)$.

Finally, if G is constant, consider the following diagram:

$$\begin{array}{ccccccccc} 0 & \to & \underline{S}_{\Phi_U}(U,G) & \to & \underline{S}_{\Phi}(X,G) & \to & \underline{S}_{\Phi}(X,G)/\underline{S}_{\Phi}(U,G) & \to & 0 \\ & & \downarrow \varphi' & & \downarrow \varphi & & \downarrow \varphi'' & & \\ 0 & \to & \Gamma_{\Phi}(S \otimes G_U) & \to & \Gamma_{\Phi}(S \otimes G) & \to & \Gamma_{\Phi}(S \otimes G_{X-U}) & \to & 0 \end{array}$$

Here φ is the usual map while φ' is the composition $S_{\Phi_U}(U,G) \overset{\varphi}{\to} \Gamma_{\Phi_U}(S^+ \otimes G) \to \Gamma_{\Phi_U}(S \mid U \otimes G) \to \Gamma_{\Phi}(S \otimes G_U)$. The map φ'' is defined by taking quotients. The diagram is easily seen to be commutative and so induces a map of homology sequences. But, φ' and φ induce isomorphism of homology. Therefore, by the 5-lemma, so does φ''.

THE DUALITY THEOREMS

We now assume X to be an n-manifold, possibly with boundary. By this, we mean a Hausdorff space such that every point has a neighborhood homeomorphic to a closed n-cell. A point is called an interior point if it has a neighborhood homeomorphic to an open n-cell. Otherwise, it is called a boundary point. The set of boundary points is denoted by $\dot{X}$.

Lemma 2: Let X be an n-manifold with or without boundary. Let S be the sheaf of singular chains over X with coefficients in K. Then $H_i(S) = 0$ for $i \neq n$, $H_n(S)_x = K$ if x is an interior point, and $H_n(S)_x = 0$ if x is a boundary point. Furthermore, $H_n(S) \mid X-\dot{X}$ is locally constant (and so, locally isomorphic to K).

Proof: If x is an interior point of X, it has an open neighborhood homeomorphic to an open n-cell. If x is a boundary point, it has an open neighborhood homeomorphic to a closed half space of euclidean n-sapce R^n.

By taking small neighborhoods of these types and applying excision, we see that it is sufficient to prove the result for the case $X = R^n$ and the case where X is a closed half space of R^n. In either case, X has a base of convex open sets. If V is such a set, then $H_i(X,X-V) = 0$ if $i \neq n$. If X is a closed half space and V meets its boundary $\dot{X}$, then $H_n(X,X-V) = 0$ also because X-V is a deformation retract of X. Finally, if V does not meet X, then $H_n(X,X-V) \approx K$. We can choose a definite isomorphism for all V by orienting X and taking the particular isomorphism determined by the orientation. This shows that $H_n(S)$ is constant over $X-\dot{X}$ and zero on X.

<u>Notation</u>: We write $T = H_n(S)$.

<u>Corollary</u>: If G is any sheaf over X, $H_n(S \otimes G) = T \otimes G$.

<u>Proof</u>: For any chain complex S, there is a natural map $H_n(S) \otimes G \to H_n(S \otimes G)$. Since $H_i(S)_x$ is either 0 or K for all i, x, and K, this map is an isomorphism. The quickest way to see this is to notice that $S = S'' \otimes_Z K$ where S" is the sheaf of singular cochains with integral coefficients. Therefore, $S \otimes_K G = S'' \otimes_Z K \otimes_K G = S'' \otimes_Z G$. This shows that we can assume the ground ring is the ring of integers Z. The result then follows immediately from the Kunneth relations.

<u>Remark</u>: If X is a manifold in the classical sense, it is easy to see that $H_n(S) \mid X-\dot{X}$ is constant if and only if X is orientable. For the general manifolds considered here, this can be taken as the definition of orientability.

<u>Remark</u>: In proving the duality theorems, we use only the following properties of n-manifolds:

1. $H_i(S \otimes G) = 0$ if $i \neq n$ while
   $H_n(S \otimes G) = T \otimes G$ for all G.
2. If Φ is PF, Φ - dim $x < \infty$.

Property (1) has just been proved. Property (2) follows from the theorem proved in the section on Čech cohomology because $A \in \Phi$ implies dim $A \leq$ dim $X < \infty$. (Theorem VIII, 1.)

<u>Theorem 1</u>: (Poincaré - Lefschetz duality).

Let X be an n-manifold. Let Φ be PF. Let G be any sheaf over X. Then there is a natural isomorphism $\alpha\colon H_i^{\Phi}(X,G) \to H_{\Phi}^{n-i}(X, T \otimes G)$.

<u>Proof</u>: Proposition 5, chapter IX shows that $\alpha\colon H^{j+(-n)}(\Gamma_{\Phi}(S \otimes G)) \to H_{\Phi}^{j}(X,\ H^{-n}(S \otimes G))$ is an isomorphism. The conditions of the proposition are satisfied because of conditions (1) and (2) in the preceding remark and the fact that $S \otimes G$ is homotopically fine. We now write $-i = j - n$ and remember that $S^k = S_{-k}$ for all k by definition.

<u>Theorem 2</u>: (Alexander-Lefschetz duality)

Let X be an n-manifold. Let Φ be PF. Let G be any sheaf over X.

Let $U \subset X$ be an open set. Let $A = X - U$. Then there is an isomorphism between the cohomology sequence of (X,A) and the homology sequence of (X,U) as follows:

$$\begin{array}{ccccccccc}
\cdots \to & H_i^{\Phi_U}(U, G \mid U) & \to & H_i^{\Phi}(X,G) & \to & H_i^{\Phi}(X,U;G) & \xrightarrow{\delta} & \cdots \\
& \approx \downarrow \alpha & & \approx \downarrow \alpha & & \approx \downarrow \alpha & & \\
\cdots \to & H_{\Phi_U}^{n-i}(U, (T \otimes G) \mid U) & \to & H_{\Phi}^{n-i}(X, T \otimes G) & \to & H_{\Phi_A}^{n-i}(A, (T \otimes G) \mid A) & \xrightarrow{\delta} & \cdots
\end{array}$$

<u>Proof</u>:

We apply proposition 6, chapter IX to the sequence

$$0 \to S \otimes G_U \to S \otimes G \to S \otimes G_A \to 0.$$

This gives a map of exact sequences as above but with $H_i^{\Phi}(\Gamma_{\Phi}(S \otimes G_U))$ in place of $H_i^{\Phi_U}(U, G \mid U)$. However, these are isomorphic by lemma 1. Now, U is also an n-manifold. We must check that the map α obtained by a substituting $H_i^{\Phi_U}(U, G \mid U)$ for $H_i^{\Phi}(\Gamma_{\Phi}(S \otimes G_U))$ agrees with the map α considered in the previous theorem. Let PQ^* be the canonical flabby resolution (over X or U). Clearly $PQ^*(G \mid U) = PQ^*(G) \mid U$ for all G, since U is open. Let $\mathfrak{F}_{\Phi} = \Gamma_{\Phi} \circ PQ^*$ and $\mathfrak{F}_{\Phi_U} = \Gamma_{\Phi_U} PQ^*$. Then $\mathfrak{F}_{\Phi_U}(G \mid U) = \mathfrak{F}_{\Phi}(G_U)$ for all G. Let S^+ be the sheaf of singular chains over U, and $S^+ \to S \mid U$ the canonical map defined in lemma 1. Let S' and $S^{+\prime}$ be the auxiliary sheaves used to define α. Then

$$\begin{array}{ccccccc}
\Gamma_{\Phi}(S \otimes G_U) & \to & \mathfrak{F}_{\Phi}(S \otimes G_U) & \leftarrow & \mathfrak{F}_{\Phi}(S' \otimes G_U) & \to & \mathfrak{F}_{\Phi}(T \otimes G_U) \\
\Vert & & \Vert & & \Vert & & \Vert \\
\Gamma_{\Phi_U}((S \otimes G) \mid U) & \to & \mathfrak{F}_{\Phi_U}((S \otimes G) \mid U) & \leftarrow & \mathfrak{F}_{\Phi_U}((S' \otimes G) \mid U) & \to & \mathfrak{F}_{\Phi_U}((T \otimes G) \mid U) \\
\uparrow & & \uparrow & & \uparrow & & \Vert \\
\Gamma_{\Phi_U}(S^+ \otimes G) & \to & \mathfrak{F}_{\Phi_U}(S^+ \otimes G) & \leftarrow & \mathfrak{F}_{\Phi_U}(S^{+\prime} \otimes G) & \to & \mathfrak{F}_{\Phi_U}((T \otimes G) \mid U)
\end{array}$$

The top row gives the map α occurring in proposition 5, chapter IX. The bottom row gives the map α of the previous theorem. The vertical map on the left gives the canonical identification $H_i(\Gamma_\Phi(S \otimes G_U)) \approx H_i^\Phi(U,G \mid U)$ of lemma 1.

The theorem is now a consequence of the previous theorem, applied to X and U, and the 5-lemma.

The classical Poincaré, Alexander, and Lefschetz duality theorems are immediate consequences of the above theorems, except for one half of the Lefschetz duality theorem which requires an additional argument.

Theorem 3: Let X be an n-manifold with boundary. Let Φ be the collection of compact sets of X. Let G be a constant sheaf. Then $H_i(X,\dot{X};G) \approx H_\Phi^{n-i}(X, T \otimes G)$ naturally.

Proof: Let $Y = X \cup \dot{X} \times I$ be the mapping cylinder of $\dot{X} \to X$. Let $U = Y - X$. Then, bu the Alexander-Lefschetz theorem, $H_i(Y,U;G) \approx H_\Phi^{n-i}(X,T \otimes G)$.

Let $(Y,U) \to (X,\dot{X})$ be the identity on X and the projection $\dot{X} \times I \to \dot{X}$ on $\dot{X} \times I$. This clearly gives homotopy equivalences $Y \to X$ and $U \to \dot{X}$. Therefore, the 5-lemma applied to the exact homology sequences of (Y,U) and $(X,\dot{X})$ shows that $H_i(Y,U;G) \approx H_i(X,\dot{X};G)$.

XII. CUP AND CAP PRODUCTS

§1.1

I will here show how to define cup products in the spectral sequences defined in Chapters IX and X. I will also give a general definition of cap-products and an application of these to the Poincaré and Alexander duality theorems. Both of these products will be defined by first defining them for natural resolutions and resolvent functors, and constructing them from a weaker kind of product, the qip[1] product. This last product is easy to derive in the case of sheaves.

Consider two exact K-categories $\mathfrak{a}, \mathfrak{B}$ admitting bilinear maps and tensor products. (cf. remarks at the beginning of chapter IV, §4: a method of discussing these concepts in general would be useful, but does not, at present, seem to be available). Let us be given three left-exact, covariant, K-functors T, T', T": $\mathfrak{a} \to \mathfrak{B}$ and assume once and for all that, for any A, B $\in \mathfrak{a}$ we have a natural map

$$\eta^{A,B}: T'(A) \otimes T''(B) \to T(A \otimes B).$$

The following notations will be used throughout this chapter:

C', C'_1, C'_2, etc., are natural T'-resolutions
C'', C''_1, C''_2, etc., " " T"- "
C, C_1, C_2, etc., " " T - "
$\mathfrak{T}', \mathfrak{T}'_1, \mathfrak{T}'_2$, etc., are resolvant functors for T'
$\mathfrak{T}'', \mathfrak{T}''_1, \mathfrak{T}''_2$, etc., " " " " T"
$\mathfrak{T}, \mathfrak{T}_1, \mathfrak{T}_2$, etc., " " " " T

<u>Definition</u>: A natural cup product from $C' \otimes C''$ to C is a cochain map $C'(A) \otimes C''(B) \to C(A \otimes B)$, natural for all A, B $\in \mathfrak{a}$, and which preserves augmentation; i.e., the

---

[1] This name is due to Miss M. Rochat.

following diagram commutes:

$$\begin{array}{ccc} & A \otimes B & \\ \epsilon' \otimes \epsilon'' \swarrow & & \searrow \epsilon \\ C'(A) \otimes C''(B) & \rightarrow & C(A \otimes B) \end{array}$$

Similarly, a natural cup product from $\mathfrak{T}' \otimes \mathfrak{T}''$ to $\mathfrak{T}$ is a natural cochain map $\mathfrak{T}'(A) \otimes \mathfrak{T}''(B) \rightarrow \mathfrak{T}(A \otimes B)$ for each $A,B \in \mathcal{a}$ which preserves augmentation; i.e., the following diagram commutes:

$$\begin{array}{ccc} T'(A) \otimes T''(B) & \xrightarrow{\eta} & T(A \otimes B) \\ \epsilon' \otimes \epsilon'' \downarrow & & \downarrow \epsilon \\ \mathfrak{T}'(A) \otimes \mathfrak{T}''(B) & \rightarrow & \mathfrak{T}(A \otimes B). \end{array}$$

Remark: $T' \circ C'$, $T'' \circ C''$ and $T \circ C$ are resolvent functors for T', T'' and T respectively (proposition III, 8.).
If $C' \otimes C'' \rightarrow C$ is a natural cup product, so clearly is the induced map $T' \circ C' \otimes T'' \circ C'' \rightarrow T \circ C$.
Remark: If $\mathfrak{T}' \otimes \mathfrak{T}'' \rightarrow \mathfrak{T}$ is a natural cup product, it induces a map $R^pT' \otimes R^qT'' \rightarrow R^{p+q}T$ which is a cup product of cohomological δ-functors (cf. chapter IV. §4). Thus this cup product agrees with any other, if we make the additional assumptions of theorem IV. 2. We leave the verification of these statements to the reader.
Definition: A natural qip product from C'' to C is a natural augmentation preserving cochain map $A \otimes C''(B) \rightarrow C(A \otimes B)$, $A, B \in \mathcal{a}$. Here, of course, A is to be regarded as a trivial cochain complex, zero except in dimension zero.

A natural qip product from $T' \otimes \mathfrak{T}''$ to $\mathfrak{T}$ is a natural, augmentation preserving, cochain map $T'(A) \otimes \mathfrak{T}''(B) \rightarrow \mathfrak{T}(A \otimes B)$, $A,B \in \mathcal{a}$.
Remark: Any natural qip product from C'' to C determines a natural qip produce from $T \otimes T'' \circ C''$ to $T \circ C$ of resolvent functors.
§1.2

We can derive cup products from certain qip products. Let C_1, C_2 be natural T-resolutions. Consider $C_3 = C_1 \circ C_2$. This is the double complex defined as follows:
$C_3^{p,q}(A) = C_1^p(C_2^q(A))$ with δ on $C_1^p(C_2^q(A))$ defined to be the

sum of δ: $C_1^p(C_2^q(A)) \to C_1^{p+1}(C_2^q(A))$ and $(-1)^p C_1^p(\delta)$: $C_1^p(C_2^q(A)) \to C_1^p(C_2^{q+1}(A))$. It is augmented by $A \xrightarrow{\epsilon} C_2^o(A) \xrightarrow{\epsilon} C_1^o(C_2^o(A))$. Consider the associated cochain complex obtained by taking $\overline{C}_3^n(A) = \sum_{p+q=n} C_3^{p,q}(A)$.

<u>Lemma 1</u>: $\overline{C}_3$ is a natural T-resolution.

<u>Proof</u>: Clearly $\overline{C}_3$ is an exact functor. $C_3^{p,q}(A)$ is T-acyclic for all p,q,A since $C_1^p(B)$ is T-acyclic for all B. We must show ϵ_*: $A \to H(C_3(A))$ is an isomorphism. Regard A as a trivial double complex, ϵ as a map of double complexes and consider the first spectral sequences of these complexes. Since C_1 is an exact functor, the E_1 term of $C_3(A)$ is $(C_1(H(C_2(A))) = C_1(A)$. Therefore, the E_2 term is $H(C_1(A)) = A$. Thus ϵ induces an isomorphism of the E_2 terms; the result follows immediately from this.

Note that the T-acyclicity of C_2 was not used.

Now let μ_1: $A \otimes C''(B) \to C_1(A \otimes B)$ and μ_2: $A \otimes C'(B) \to C_2(A \otimes B)$ be natural qip products. For convenience we re-write the second one as $C'(B) \otimes A \to C_2(B \otimes A)$. Consider

$$C'(A) \otimes C''(B) \xrightarrow{\mu} C_1(C'(A) \otimes B) \xrightarrow{C_1(\mu_2)} C_1(C_2(A \otimes B)).$$

This is clearly a natural, augmentation preserving, cochain map. By lemma 1, $C_1 \circ C_2$ is a natural T-resolution. Thus we have constructed a natural cup product.

<u>§1.3</u>

We now consider products involving resolvant functors. A map $(\mathfrak{F}_1', \mathfrak{F}''_1, \mathfrak{F}_1) \to (\mathfrak{F}_2', \mathfrak{F}_2'', \mathfrak{F}_2)$ (using notation as described above) shall be a set of three maps $\mathfrak{F}_1' \to \mathfrak{F}_2'$, $\mathfrak{F}_1'' \to \mathfrak{F}_2''$ and $\mathfrak{F}_1 \to \mathfrak{F}_2$. If we have natural cup products, it is clear what we mean by saying that such a map preserves cup products. Similarly, if we have natural qip products, it is clear what we mean by saying that a map $(\mathfrak{F}_1'', \mathfrak{F}_1) \to (\mathfrak{F}_1'', \mathfrak{F}_2)$ preserves these products.

Suppose that $\mathfrak{F}' \otimes \mathfrak{F}'' \xrightarrow{\mu} \mathfrak{F}$ and $C' \otimes C'' \xrightarrow{\tau} C$ are natural cup products. By proposition IX. 3 $\mathfrak{F}' \circ C'$, $\mathfrak{F}'' \circ C''$ and $\mathfrak{F} \circ C$ are again resolvant functors. We can define a natural cup product $\mathfrak{F}' \circ C' \otimes \mathfrak{F}'' \circ C'' \to \mathfrak{F} \circ C$ to be the composition

$$\mathfrak{F}'(C'(A)) \otimes \mathfrak{F}''(C''(B)) \xrightarrow{\mu} \mathfrak{F}(C'(A) \otimes C''(B)) \xrightarrow{\mathfrak{F}(\tau)} \mathfrak{F}(C(A \otimes B))$$

Clearly, the augmentations

$$(\mathcal{F}'(A), \mathcal{F}''(B), \mathcal{F}(A \otimes B)) \xrightarrow{\mathcal{F}(\epsilon)} (\mathcal{F}'(C'(A)).\mathcal{F}''(C''(B)), \mathcal{F}(C(A \otimes B)))$$

and

$$(T'(C'(A)), T''(C''(B)), T(C(A \otimes B)) \xrightarrow{\epsilon} (\mathcal{F}'(C'(A)), \mathcal{F}''(C''(B)), \mathcal{F}(C(A \otimes B))$$

preserve these products.

An examination of the proof of the corollary of proposition IX. 3 now yields the following result.

Lemma 2: Let $\mathcal{F}'_0 \otimes \mathcal{F}''_0 \to \mathcal{F}_0$ and $\mathcal{F}'_1 \otimes \mathcal{F}''_1 \to \mathcal{F}_1$ be any natural cup products. Then there are natural cup products $\mathcal{F}'_n \otimes \mathcal{F}''_n \to \mathcal{F}_n$ for $n = 2,3,4$ and product preserving maps as follows:

$$(\mathcal{F}'_0, \mathcal{F}''_0, \mathcal{F}_0) \to (\mathcal{F}'_2, \mathcal{F}''_2, \mathcal{F}_2) \leftarrow (\mathcal{F}'_3, \mathcal{F}''_3, \mathcal{F}_3) \to (\mathcal{F}'_4, \mathcal{F}''_4, \mathcal{F}_4) \leftarrow (\mathcal{F}'_1, \mathcal{F}''_1, \mathcal{F}_1)$$

Proof: We have only to choose a natural cup product $C' \otimes C'' \to C$, define the $\mathcal{F}_n$, etc., as in the proof of the corollary to proposition IX. 3 and define the products $\mathcal{F}'_n \otimes \mathcal{F}''_n \to \mathcal{F}_n$ as indicated in the discussion preceding this lemma.

Remark: A similar result holds for natural qip products. The statement and proof are the same except that all $\mathcal{F}'_n$ are replaced by T' and C' is replaced by the identity functor.

§1.4

It is now very easy to define cup products in the spectral sequences. Suppose $M, N \in \mathcal{C}(\mathcal{A})$. Let $\mathcal{F}' \otimes \mathcal{F}'' \to \mathcal{F}$ be a natural cup product. Then $\mathcal{F}'(M) \otimes \mathcal{F}''(N) \to \mathcal{F}(M \otimes N)$ is a map of double complexes. By CE chapter XV Ex. 1,2, and 4, we get a product from the spectral sequences of M and N to the spectral sequences of $M \otimes N$; this product is clearly natural. To show it is independent of $\mathcal{F}'$, $\mathcal{F}''$, $\mathcal{F}$ and $\mathcal{F}' \otimes \mathcal{F}'' \to \mathcal{F}$ we note that by lemma 2 it is sufficient to consider the case where we have a product preserving map $(\mathcal{F}'_0, \mathcal{F}''_0, \mathcal{F}_0) \to (\mathcal{F}'_1, \mathcal{F}''_1, \mathcal{F}_1)$. But in this case it is clear that the resulting isomorphism of spectral sequences preserves products.

In chapter IX, we saw that $E_2^{p,q} = R^pT'.H^q(M)$ and that $E_\infty^{p,q}$ is often isomorphic to $\mathcal{G}^{p,q}H^{p+q}(T'(M))$ with similar results for the spectral sequences of N and $M \otimes N$. It is easy to verify that these isomorphisms preserve products if the product $H^r(T'(M)) \otimes H^s(T''(N)) \to H^{r+s}(T(M \otimes N))$ is induced by η: $T' \otimes T'' \to T$, and the product

$$R^pT'.H^q(M) \otimes R^rT''.H^s(N) \to R^{p+r}T.H^{q+s}(M \otimes N)$$

is defined to be the composition

$$R^pT'.H^q(M) \otimes R^rT''.H^s(N) \xrightarrow[\text{product}]{\text{cup}} R^{p+r}T.H^q(M) \otimes H^s(N) \xrightarrow{(-1)^{rq}\tau_*} R^{p+r}T.H^{q+s}(M \otimes$$

where τ: $H^q(M) \otimes H^s(N) \to H^{q+s}(M \otimes N)$ is the usual map of the Kunneth formulas.

§2 Cap products

Let $M \in \mathcal{C}(\mathcal{a})$ and $A \in \mathcal{a}$. Assume $HR^pT.(M \otimes A) = 0$ for $p \neq 0$. Then ϵ_*: $HT(M \otimes A) \to H\mathfrak{F}(M \otimes A)$ is an isomorphism. We now define the cap product

$$H^p(T'(M)) \otimes R^qT''.(A) \to H^{p+q}(T(M \otimes A))$$

to be the composition of ϵ_*^{-1} with the cohomology induced by a natural qip product $T'(M) \otimes \mathfrak{F}''(A) \to \mathfrak{F}(M \otimes A)$. The remark following lemma 2 shows that this product is independent of the choice of $\mathfrak{F}''$, $\mathfrak{F}$, and the natural qip product. (This need only be verified when we have a product preserving map $(\mathfrak{F}''_o, \mathfrak{F}_o) \to (\mathfrak{F}''_1, \mathfrak{F}_1)$).

A fundamental property of this product is given by the following lemma, which relates cap products to cup products and the map α defined in chapter IX.

<u>Lemma 3</u>: Suppose that M is bounded below or that T-dim $\mathcal{a}$ and T'-dim $\mathcal{a}$ are finite. Assume $H^q(M)$ and $H^q(M \otimes A)$ are zero for $q > k$. Thus the maps α are defined in the diagram below. Assume also that $HR^pT.(M \otimes A) = 0$ for $p \neq 0$, so that cap products are defined. Then the following diagram commutes:

$$\begin{array}{ccc} H^{p+k}(T'(M)) \otimes R^qT''.(A) & \xrightarrow{\text{cap product}} & H^{p+q+k}(T(M \otimes A)) \\ \alpha \otimes 1 \downarrow & & \downarrow \alpha \\ R^pT'.H^k(M) \otimes R^qT''.(A) & \longrightarrow & R^{p+q}T.H^k(M \otimes A) \end{array}$$

Here the bottom map is cup product followed by the map induced by the usual map $H^k(M) \otimes A \to H^k(M \otimes A)$.

<u>Proof</u>: Choose a natural cup product μ: $\mathfrak{F}' \otimes \mathfrak{F}'' \to \mathfrak{F}$. Define the cap product using the natural qip product given by $T' \otimes \mathfrak{F}'' \xrightarrow{\epsilon \otimes 1} \mathfrak{F}' \otimes \mathfrak{F}'' \xrightarrow{\mu} \mathfrak{F}$. Let M_1 be the complex used to define α; i.e., M_1^q is M^q if $q \leq k$, $Z^k(M)$ if $q = k$, and 0 if $p > k$. Let $(M \otimes A)_1$ be the analagous complex

obtained from $M \otimes A$. Clearly there is a natural map $M_1 \otimes A \to (M \otimes A)_1$ such that the diagram

$$\begin{array}{ccccc} M \otimes A & \leftarrow & M_1 \otimes A & \xrightarrow{\gamma \otimes 1} & H^k(M) \otimes A \\ & \nwarrow & \downarrow & & \downarrow \\ & & (M \otimes A)_1 & \xrightarrow{\gamma} & H^k(M \otimes A) \end{array}$$

commutes. γ is the map used in defining α on p.

Using this and the particular choice of the natural qip product, we see immediately that the following diagram commutes

$$\begin{array}{ccccccccc} T'(M) \otimes \mathfrak{F}''(A) & \xrightarrow{\epsilon \otimes 1} & \mathfrak{F}'(M) \otimes \mathfrak{F}''(A) & \leftarrow & \mathfrak{F}'(M_1) \otimes \mathfrak{F}''(A) & \rightarrow & \mathfrak{F}'(H^k(M)) \otimes \mathfrak{F}''(A) \\ \text{qip} \downarrow & & \downarrow \text{cup} & & \downarrow \text{cup} & & \downarrow \text{cup} \\ T(M \otimes A) \xrightarrow{\epsilon} \mathfrak{F}(M \otimes A) & \xrightarrow{=} & \mathfrak{F}(M \otimes A) & \leftarrow & \mathfrak{F}(M_1 \otimes A) & \rightarrow & \mathfrak{F}(H^k(M) \otimes A) \\ & & & \nwarrow & \downarrow & & \downarrow \\ & & & & \mathfrak{F}((M \otimes A)_1) & \rightarrow & \mathfrak{F}(H^k(M \otimes A)) \end{array}$$

We now pass to cohomology. The top row defines the map $\alpha \otimes 1$, while the composition of the lower maps defines α for $M \otimes A$. The result follows immediately from the commutativity of the diagram.

Corollary: Suppose the conditions of the lemma are satisfied and, in addition, that

(a) $H^q(M) = 0$ and $H^q(M \otimes A) = 0$ for $q \neq 0$, and

(b) $HR^pT.(M) = 0$ for $p \neq 0$

Then α: $H^p(T'(M)) \to R^pT'.H^o(M)$ and α: $H^p(T(M \otimes A)) \to R^pT.H^o(M \otimes A)$ are both isomorphisms (proposition IX.5); under these isomorphisms cap product corresponds to cup product.

§3.1

We will now show how the preceding theory applies to sheaves. Let $\mathcal{A}$ be the category of sheaves on X, $\mathcal{B}$ that of modules. Let Φ, Φ', Φ'' be families of supports in X such that $\Phi \supset \Phi' \cap \Phi''$. (The theory of cup products in chapter IV §4.2 carries over to this more general case; there we assumed $\Phi = \Phi' \cap \Phi''$.) Then there is a natural map from $\Gamma_{\Phi'} \otimes \Gamma_{\Phi''} \to \Gamma_{\Phi}$. We keep the notations of §1 with $T' = \Gamma_{\Phi'}$. $T'' = \Gamma_{\Phi''}$ and $T = \Gamma_{\Phi}$; of course $R^*T' = H^*_{\Phi'}(X, \ )$, etc.

We show the existence of a natural qip product from the canonical flabby resolution to itself, and this will show the existence of the other products. Recall that $P(F) = \tilde{\tilde{F}}$ and $Q(F) = P(F)/F$. For all $F, G \in \mathcal{A}$, there is a natural isomorphism

$\overline{F} \otimes \overline{G} \to \overline{F \otimes G}$. Apply L to the composition $\Gamma(,\overline{F}) \otimes \Gamma(,G) \stackrel{\approx}{\to} \Gamma(,\overline{F} \otimes \overline{G}) \to \Gamma(,\overline{F \otimes G})$, where the first map is the obvious natural isomorphism, and we obtain a natural map $P(F) \otimes P(G) \to P(F \otimes G)$.

The following diagram obviously commutes:

$$\begin{array}{ccc} & F \otimes G & \\ i \otimes i \swarrow & & \searrow i \\ P(F) \otimes P(G) & \to & P(F \otimes G) \end{array}$$

Here i is the usual inclusion map $F \to P(F)$. We now define a map $f\colon F \otimes P(G) \to P(F \otimes G)$ to be the composition $F \otimes P(G) \xrightarrow{i \otimes 1} P(F) \otimes P(G) \to P(F \otimes G)$. Consider the diagram:

$$\begin{array}{ccccccccc} 0 \to & F \otimes G & \to & F \otimes P(G) & \to & F \otimes Q(G) & \to 0 \\ & \downarrow = & & \downarrow f & & \downarrow g & \\ 0 \to & F \otimes G & \to & P(F \otimes G) & \to & Q(F \otimes G) & \to 0 \end{array}$$

The top sequence is exact since $0 \to G \to P(G) \to Q(G) \to 0$ is algebraically split. The first square commutes, and the bottom sequence is also exact. Therefore (cf. lemma III. 1) there is a unique map g making the diagram commutative. We now define maps $F \otimes PQ^i(G) \to PQ^i(F \otimes G)$ for all i by using f and g as follows:

$$F \otimes PQ^i(G) \xrightarrow{f} P(F \otimes Q^i(G)) \xrightarrow{P(g)} P(Q(F \otimes Q^{i-1}(G))) \xrightarrow{PQ(g)} \cdots \to PQ^i(F \otimes$$

These compositions give us a map $F \otimes PQ^*(G) \to PQ^*(F \otimes G)$; the reader will easily verify that this is a natural qip product.

The earlier theory now enables us to construct cup products $PQ^* \otimes PQ^* \to PQ^*$ and $\Gamma_{\Phi}\circ PQ^* \otimes \Gamma_{\Phi''}\circ PQ^* \to \Gamma_{\Phi}\circ PQ^*$ and, under certain conditions, cap products

$$H^p(\Gamma_{\Phi'}(M)) \otimes H^q_{\Phi''}(X_1 F) \to H^{p+q}(\Gamma_{\Phi}(M \otimes A))$$

where M is a cochain complex of sheaves. We need here a map $PQ^* \circ PQ^* \to PQ^*$. This can be constructed since $PQ^* \circ PQ^*$ is easily seen to be naturally algebraically split.

The image of $x \otimes y$ under cup product is denoted by $x \cup y$, of $a \otimes b$ under cap product by $a \cap b$, where a,b,x,y are elements of suitable objects.

§3.2

We now define cup products in the spectral sequence of a map. This requires a slightly more general definition of this sequence. The notation will be as in chapter X. The map is $f: E \to B$, G is a sheaf over E and I is an injective resolution of G. The spectral sequence of f is defined to be the first spectral sequence of $f_\Psi(I)$. We must show that the injective resolution can be replaced by any flabby resolution C of G. Now there is a map $C \to I$ unique up to homotopy; this induces a unique map from the spectral sequence of $f_\Psi(C)$ to that of $f_\Psi(I)$, and we must show that this is an isomorphism. The E_2 terms are $H_\Phi(B,H(f_\Psi(C)))$ and $H_\Phi(B,H(f_\Psi(I)))$. Therefore, it is sufficient to show that $H(f_\Psi(C)) \to H(f_\Psi(I))$ is an isomorphism. But $f_\Psi(F)$ is defined by the stack $\Gamma_{\Psi(U)}(f^{-1}(U),F) = \Gamma_{\Psi(U)}(F \mid f^{-1}(U))$. Since $C \mid f^{-1}(U)$ and $I \mid f^{-1}(U)$ are both flabby resolutions of $G \mid f^{-1}(U)$ it follows from proposition III. 7 that $H(\Gamma_{\Psi(U)}(C \mid f^{-1}(U))) \to H(\Gamma_{\Psi(U)}(I \mid f^{-1}(U)))$ is an isomorphism. The desired result now follows by applying L.

Suppose now that we have three systems of supports $(\Phi_1, \Psi_1(U))$, $(\Phi_2, \Psi_2(U))$, $(\Phi, \Psi(U))$ each satisfying the conditions of chapter X, p. 126. Suppose also that $\Phi_1 \cap \Phi_2 \subset \Phi$ and $\Psi_1(U) \cap \Psi_2(U) \subset \Psi(U)$ for all U. If G_1 and G_2 are sheaves over E the usual map

$$\Gamma_{\Psi_{1}(U)}(f^{-1}(U),G_1) \otimes \Gamma_{\Psi_{2}(U)}(f^{-1}(U),G_2) \to \Gamma_{\Psi(U)}(f^{-1}(U),G_1 \otimes G_2)$$

induces a natural map $f_{\Psi_1}(G_1) \otimes f_{\Psi_2}(G_2) \to f_\Psi(G_1 \otimes G_2)$.

Now, let PQ^* denote the canonical flabby resolution, and let I be an injective resolution of $G_1 \otimes G_2$. Then $PQ^*(G_1) \otimes PQ^*(G_2)$ is an acyclic complex over $G_1 \otimes G_2$, since PQ^* is algebraically split. Consequently, there is an augmentation preserving cochain map $PQ^*(G_1) \otimes PQ^*(G_2) \to I$, and this map is unique up to homotopy. The composition

$$f_{\Psi_1}(PQ^*(G_1)) \otimes f_{\Psi_2}(PQ^*(G_2)) \to f_\Psi(PQ^*(G_1) \otimes PQ^*(G_2)) \to f_\Psi(I)$$

then gives the required product from the spectral sequences of Ψ_1, G_1 and Ψ_2, G_2 to that of Ψ, $G_1 \otimes G_2$.

In chapter X we have seen that under certain conditions, $E_\infty^{i,j} \approx \mathcal{G} H_{\Psi''}^{i+j}(E,G)$. This isomorphism is given by the augmentation and the fact that $\Gamma_\Phi \circ f_\Psi = \Gamma_{\Psi''}$. Clearly, the produce in E_∞ corresponds, under this isomorphism, to the ordinary cup produce in $H^{i+j}_{\Psi''}(E,G)$. This follows immediately from the definition of cup products in chapter IV. Similarly, we see that the product on the E_2 terms is induced by the map $R^i f_{\Psi_1}\cdot(G_1) \otimes R^j f_{\Psi_2}\cdot(G_2) \to R^{i+j} f_\Psi\cdot(G_1 \otimes G_2)$ which is in turn induced by the map of stacks given by the ordinary cup product

$$H^i_{\Psi_{1(U)}}(f^{-1}(U), G_1 \mid f^{-1}(U)) \otimes H^j_{\Psi_{2(U)}}(f^{-1}(U), G_2 \mid f^{-1}(U)) \to$$
$$H^{i+j}_{\Psi(U)}(f^{-1}(U). G_1 \otimes G_2 \mid f^{-1}(U))$$

§4

Cap products, and lemma 3, have some interesting applications. Let M be the sheaf S of singular chains, or, more generally, take $M = S \otimes F$ where F is an ordinary sheaf. If Φ is PF, then the cap product is defined, since $M \otimes G = S \otimes F \otimes G$ is homotopically fine. The cap product is a map

$$H^{\Phi'}_n(X,F) \otimes H^p_{\Phi''}(X,G) \to H^\Phi_{n-p}(X,F \otimes G).$$

Suppose now that X is a paracompact n-manifold. Let Φ' be the family of all closed subsets of X. Let $\mu \in H^{\Phi'}_n(X,K)$ be such that $\alpha(\mu) = 1 \in H^0_{\Phi'}(X,K)$.

Proposition 1: Let X be a paracompact n-manifold. Let Φ' be the family of all closed subsets of X and let μ be the element just defined. Let $\Phi'' = \Phi$ be PF. Then, for all $u \in H^p_\Phi(X,G \otimes T)$, we have $\alpha(\mu \cap u) = u$.

Proof: We merely apply the lemma 3, using the fact that $1 \cup u = u$.

Corollary: The cap product with μ gives a map $H^p_\Phi(X, G \otimes T) \to H^\Phi_{n-p}(X,G)$ which is inverse to α and hence is the canonical isomorphism of the Poincaré-Lefschetz duality theorem.

In order to get a similar result for the Alexander-Lefschetz duality theorem, we must generalize slightly the definition of cap product. In this definition, Φ is assumed to be PF. Let $U \subset X$ be open and let $A = X-U$. Consider the sequence $0 \to S_U \otimes F \to S \otimes F \to S_A \otimes F \to 0$. Let $\overline{\Gamma}_\Phi(S_A \otimes F)$ be the image of $\Gamma_\Phi(S \otimes F) \to \Gamma_\Phi(S_A \otimes F)$. Therefore, the sequence

$$0 \to \Gamma_\Phi(S_U \otimes F) \to \Gamma_\Phi(S \otimes F) \to \overline{\Gamma}_\Phi(S_A \otimes F) \to 0$$

is exact. If $\mathfrak{F}$ is any resolvent functor for Γ_Φ, we have the commutative diagram

$$\begin{array}{ccccccccc} 0 \to & \Gamma_\Phi(S_U \otimes F) & \to & \Gamma_\Phi(S \otimes F) & \to & \overline{\Gamma}_\Phi(S_A \otimes F) & \to 0 \\ & \downarrow \epsilon & & \downarrow \epsilon & & \downarrow \epsilon' & \\ 0 \to & \mathfrak{F}(S_U \otimes F) & \to & \mathfrak{F}(S \otimes F) & \to & \mathfrak{F}(S_A \otimes F) & \to 0 \end{array}$$

where ϵ' is the composition of ϵ with the inclusion $\overline{\Gamma}_\Phi \subset \Gamma_\Phi$. Since $S_U \otimes F$ and $S \otimes F$ are homotopically fine, both ϵ's induce isomorphisms of cohomology. By applying the 5-lemma to the cohomology sequences we see that ϵ' also induces an isomorphism of cohomology. Consequently we can define a cap product

$$H_n^{\Phi'}(X,U;F) \otimes H^p_{\Phi''}(X,G) \to H^\Phi_{n-p}(X,U;\ F \otimes G)$$

by comsidering the diagram

$$\overline{\Gamma}_{\Phi'}(S_A \otimes F) \otimes \mathfrak{F}''(G) \to \mathfrak{F}(S_A \otimes F \otimes G) \xleftarrow{\epsilon'} \overline{\Gamma}_\Phi(S_A \otimes F \otimes G)$$

(As usual, maps involving $\overline{\Gamma}$ are defined by first injecting it into Γ and then applying the usual maps.)

By considering the diagram

$$\begin{array}{ccc} H(\overline{\Gamma}_{\Phi'}(S_A \otimes F)) \otimes H_{\Phi''}(X,G) & \to & H(\overline{\Gamma}_\Phi(S_A \otimes F \otimes G)) \\ \downarrow & & \downarrow \\ H(\Gamma_{\Phi'}(S_A \otimes F)) \otimes H_{\Phi''}(X,G) & \to & H(\Gamma_\Phi(S_A \otimes F \otimes G)) \\ \downarrow \alpha \otimes 1 & & \downarrow \\ H_{\Phi'}(X,T \otimes F) \otimes H_{\Phi''}(X,G) & \to & H_\Phi(X,T \otimes F \otimes G) \end{array}$$

we see that lemma 3 applies also to this generalized cap product.

Let X be a paracompact n-manifold. Let Φ' be the family of all closed subsets of X. Let $\mu \in H_n^{\Phi'}(X,U;K)$ be such that $\alpha'(\mu) = 1 \in H^0_{\Phi'}(X,K)$.

<u>Proposition 2</u>: Let X be a paracompact n-manifold. Let Φ' be the family of all closed subsets of X, and let μ be the element just defined. Let Φ be PF. Then, for all $u \in H^p_\Phi(X,G \otimes T_A)$, we have $\alpha(\mu \frown U) = U$.

<u>Corollary</u>: The cap product with μ gives a map $H^p_\Phi(X,G \otimes T_A) \to H^\Phi_{n-p}(X,U;G)$ which is inverse to α' and hence is the canonical isomorphism of the Alexander-Lefschetz duality theorem.

BIBLIOGRAPHY

1. S. S. Chern, "Scientific Report of The Second Summer Institute II Complex Manifolds." Bull. A.M.S. 62, 102-117.

2. O. Zariski, "Scientific Report on The Second Summer Institute III Algebraic Sheaf Theory." Bull. A.M.S. 62, 117-141.

3. D. M. Kan, "Adjoint Functors". Trans. A.M.S. 87, 294-329 (1958).

4. B. Eckmann and A. Schöpf, "Über injective Moduln." Arch. Math 4, 75-78 (1953).

5. D. A. Buchsbaum, "Exact Categories and Duality." Trans. A.M.S. 80, 1-34 (1955).

6. F. Hirzebruch, "Arithmetic Genera and the Theorem of Riemann-Roch for algebraic varieties." Proc. Nat. Acad. Sci. U.S.A. 39, 951-956 (1953).

7. J. P. Serre, "Faisceaux Algebriques coherents." Ann. of Math. 61, (1955).